AF541093

Handbook of HONEY

Handbook of HONEY

– Author –

Revanth Bollapragada

Nagarajan Bhupatiraju

ISBN: 9789389183856 (Hardback)

Published by : **BIO-GREEN BOOKS**
4736/23, Ansari Road, Darya Ganj,
New Delhi – 110 002
Phone: 011-23245578, 23244987
E-mail: biogreenbooks@gmail.com

Printed at : **Replika Press Pvt. Ltd.**

Preface

Honey is a natural and sweet substance that is produced by honeybees from the nectar of flowers. It is a versatile and highly nutritious food that has been used for centuries for its medicinal and culinary properties. The use of honey dates back to ancient times, where it was highly valued for its medicinal properties. Honey was used to treat wounds, burns, and other skin conditions due to its antibacterial and anti-inflammatory properties. It was also used to treat digestive issues, sore throats, and coughs. The ancient Greeks and Egyptians recognized the importance of honey in their diets and used it in many of their culinary dishes. Honey is composed of simple sugars, such as glucose and fructose, as well as water and a variety of minerals, vitamins, and antioxidants. The exact composition of honey can vary depending on the floral source, climate, and beekeeping practices used to produce it. The color and flavor of honey can also vary depending on these factors.

One of the most important features of honey is its health benefits. Honey has antibacterial and anti-inflammatory properties that make it an effective natural remedy for wound healing, cough suppression, and sore throat relief. It can also help improve digestive health and boost athletic performance. Honey contains antioxidants that can help protect the body against oxidative stress and reduce the risk of chronic diseases, such as heart disease and cancer. In addition to its health benefits, honey is also a versatile ingredient in the kitchen. It can be used as a natural sweetener in baking and cooking, and can add flavor and complexity to a variety of dishes. Honey can also be used in skincare products and hair treatments due to its moisturizing and nourishing properties. There are many different types of honey available, each with its own unique flavor and nutritional profile. Some popular types include clover honey, wildflower honey, and Manuka honey. Honey can be purchased in its raw, unprocessed form or can be pasteurized and filtered for a smoother texture. Honey is a delicious and nutritious food that has been enjoyed for centuries. Its health benefits and culinary versatility make it a popular choice for both cooking and natural health remedies. With its rich history, complex flavor profile, and numerous health benefits, honey is truly a remarkable and valuable food.

It is a comprehensive guide to the physicochemical characterization, botanical and therapeutic properties, and health benefits of honey. Describes the physicochemical properties of honey, including its sugar composition, moisture content, and acidity and also discusses the botanical properties of honey derived from stingless bees. It delves into the unique features of this type of honey, including its flavor profile, nutritional content, and medicinal properties. Focuses on the various therapeutic properties of honey, including its antibacterial, anti-inflammatory, and antioxidant properties and also explores the use of honey in wound healing, cough suppression, and digestive health. The health benefits of honey, including its potential to improve cardiovascular health, lower blood sugar levels, promote weight loss in improving athletic performance and reducing allergy symptoms. This book is an essential resource for Food scientists and technologists, Beekeepers Health professionals, Researchers and academics, Regulatory bodies and policymakers who are responsible for ensuring the safety and quality of honey, and who need to be aware of the various analytical methods used to classify and authenticate it.

We are grateful to all those persons as well as various books, manuals, periodicals, magazines, journals etc. that helped in the preparation of this book. In spite of the best efforts, it is possible that some errors may have occurred into the compilation and editing of the book. Further queries, constructive suggestions and criticisms for the improvement of the book are always welcomed and shall be thankfully acknowledged.

Revanth Bollapragada

Nagarajan Bhupatiraju

Contents

Preface .... *v*

1. **Introductory Chapter: The Importance of the Physicochemical Characterization of Honey** **1**
    - Apiculture and Meliponiculture
    - Honey properties
    - References
2. **Pot-Pollen and Pot-Honey from Stingless Bees of the Alto Balsas, Michoacán, Mexico: Botanical and Physicochemical Characteristics** **5**
    - Introduction
    - Study area
    - Collection and analysis of pot-honey and pot-pollen samples
    - Pot-pollen and pot-honey characteristics
    - Conclusions
    - References
3. **Therapeutic Properties of Honey** **17**
    - Introduction
    - Medicinal properties of honey
    - Factors that may affect the therapeutic potentials of honey
    - Conclusions
    - References
4. **Health Benefits of Honey** **39**
    - Introduction
    - Properties of honey
    - Conclusions
    - References

5. A Review on Analytical Methods for Honey Classification, Identification and Authentication 53

- Introduction
- Authenticity of honey
- Detection methods and technologies
- Conclusion
- References

Index 85

Chapter 1

Introductory Chapter: The Importance of the Physicochemical Characterization of Honey

1. Apiculture and Meliponiculture

Beekeeping is an emerging activity on small farms around the world, as well as in regions where diversification of food products is essential for generating income and subsistence for families. Meliponicultural and beekeeping activities, when well managed, are profitable, with low environmental impact and require little input. But in addition to the products of the colony, there are still the benefits of cross-pollination performed by these insects, considered the most important among all animals. In many rural properties, there are inexhaustible sources of food for bees, and the pollination process carried out by them benefits agricultural crops, generating higher yields of fruits and seeds.

Among the products of the colony, honey occupies a prominent place in production, mainly because it is a valuable source of food and much appreciated due to its sweet aroma and flavor. It is estimated that in the world, there are more than 20,000 species of bees [1], but the bee *Apis mellifera* L. is the main producer of honey for human consumption due to its ancient keeping them in the main consuming countries. However, there is a great diversity of other bee species that produce honey of excellent quality and extremely appreciated, such as that produced by meliponine bees. In Brazil, there are more than 250 known species, which present heterogeneity in color, size, shape, nesting habits, and nest population [2].

Meliponine bees produce less honey compared to *A. mellifera* L., but their slightly more acidic taste, greater fluidity, and lower viscosity have attracted the most demanding consumers and are gaining more and more space in the consumer market worldwide and in haute cuisine. This linked to the higher market value of honey from meliponine bees compared to that of *A. mellifera* L. has also aroused interest in keeping them by beekeepers.

2. Honey properties

The floral honey produced by the bees originates from the nectar of the flowers and the extra floral honey from aphide excretion, after collection; they transform it using specific substances, then store, and let it mature in the colony, in pots of honey (stingless bees) or combs (*A. mellifera* L.). This product is a complex blend of nutrients, which include carbohydrates, amino acids, fatty acids, enzymes, and minerals. These substances vary within a range of minimum and maximum values; some are present in a greater proportion, but the point is that they are influenced by several factors, such as the botanical, geographical origin, and species of bee among others [3]. In addition to being a food product, honey has several benefits for human health and has for many years been used in alternative medicine. Recently, many studies have reported the effectiveness of honey for various medicinal purposes, due to its components and its antibacterial, anti-inflammatory, antioxidant, antiviral, antifungal, and anticancer properties. A particular chapter in this book will address the beneficial properties of honey for human health and factors that can alter the therapeutic properties of honey, such as the physical factors of the environment.

The expanding world honey market has intensified efforts to authenticate and characterize honey, as they play an important role for both consumers and producers. The authenticity of honey is defined internationally by the Codex Alimentarius [4], which establishes the identity and essential quality requirements of honey intended for human consumption. These standards are applied to honey produced by bees and cover all styles of honey presentations, which are processed and ultimately intended for human consumption. Studies for the authentication of honey involve various analysis techniques in order to determine the botanical and geographical origin of honey, as well as of unauthorized substances.

Honey can be called unifloral or multifloral depending on the percentage of specific pollen types present in its composition. The richness and diversity of bee or honey flora, both from wild and cultivated plants, can give rise to a variety of honey with different properties. Many studies seek to identify specific chemical markers for unifloral honey based on the analysis of data on the composition of volatile compounds, phenolic acids, flavonoids, carbohydrates, amino acids, and some other constituents of honey. However, the identification of reliable chemical markers for the discrimination of honey collected from different floral resources is still difficult due to the chemical composition of honey also depending on other factors, such as geographical origin, harvest season, storage method, bee species, and even interactions between chemical compounds and honey enzymes (Kaškonienė; Venskutonis, [5]).

In addition, the results of the honey's chemical constituents may depend on sample preparation and analysis techniques. Traditionally, physicochemical and melissopalynological analyses have been the most used to determine the botanical origin of honey. However, these techniques when performed individually can provide ambiguous results, making it difficult to discriminate between uni- and multifloral honey. Estevinho et al. [6] analyzed 112 samples of unifloral honey from *Lavandula* spp. and reported that the combination of melissopalynological and physicochemical analyses of honey associated with multivariate data processing techniques can be effective for the discrimination of uni- and multifloral honey. In this book there is a literature review that provides an overview and summary of the instrumental and analytical methods available for authenticating honey, from conventional molecular techniques to the most recent ones, being very useful as a guide for choosing the appropriate method for analysis, classification, and honey authentication.

The global concern is to carry out more and more a characterization of regional honey to strengthen local markets, such as this one from a region of Mexico, that presents the advances in the characterization of botanical origin of stingless bees' honey, and the analysis of their physicochemical properties in the Alto Balsas, Michoacan, Mexico, or this which aims at botanical characterization of *Apis mellifera* honey samples from the main beekeeping Mexican regions with melissopalynological studies. This method has been used with more intensity because it allows to know the plants visited by the bees and to identify the pollen grains present in honey. In addition, some countries do not have standard regulatory norms for honey from stingless bees or even honeybees. Others need to update these standards, so we present in this book more data about that from countries like Uganda where several samples do not fit the country's standards.

It is hoped that this book will help in the discussion on the identification/location of honey, as well as in the creation and/or updating of standard norms at the national and even international level.

References

[1] Michener CD. The bees of the world. Baltimore: Johns Hopkins University Press; 2000. 913p

[2] Pedro SRM. The stingless bee fauna in Brazil (Hymenoptera: Apidae). Sociobiology. 2014;**6**:348-354

[3] Araujo JS, Chambo ED, Costa MAPC, Silva SMPC, Carvalho CAL, Estevinho MLM. Chemical composition and biological activities of mono- and heterofloral bee pollen of different geographical origins. International Journal of Molecular Sciences. 2017;**18**:921

[4] CODEX ALIMENTARIUS. Revised codex standard for honey. Codex STAN 12-1981; Rev. 1, 1987. Rev. 2, 2001

[5] Kaškonienė V, Venskutonis PR. Floral markers in honey of various botanical and geographic origins: A review. Comprehensive Reviews in Food Science and Food Safety. 2010;**9**:620-634

[6] Estevinho LM, Chambó ED, Pereira APR, Carvalho CAL, Toledo VAA. Characterization of *Lavandula* spp. honey using multivariate techniques. PLoS ONE. 2016;**11**

Chapter 2

Pot-Pollen and Pot-Honey from Stingless Bees of the Alto Balsas, Michoacán, Mexico: Botanical and Physicochemical Characteristics

Abstract

The demand for stingless bees' products (pot-honey and pot-pollen) has increased. No formal quality standards have been defined, which is very complex, because of the variety of species and types of honey specific to each region. For this reason, it is important to deepen the understanding of stingless bees' honey characteristics. From the above, the aim of this chapter is to present the advances in the characterization of botanical origin of stingless bees' honey, and the analysis of their physicochemical properties in the Alto Balsas, Michoacán, Mexico, as a way to contribute to the strengthening of new local economic strategies, generating information on the quality of the honey produced in the region.

Keywords: Meliponini, meliponiculture, botanical origin of honey, melissopalynological analysis, honey quality, analytical electrochemistry

1. Introduction

The work we present is part of a research project that has been carried out since 2014 to date, in the area that we refer to as *Alto Balsas Michoacano*. The main objective of the project has been to generate information about the human-stingless bees system in the perspective of contributing to the construction of a harmonious and sustainable relationship between the inhabitants of the study region, their surrounding ecosystems and this group of insects. This has allowed us to orient different lines of research to generate a comprehensive understanding of the relationship of humans with stingless bees in this particular region. Specifically we have directed our efforts to: (1) document the diversity of stingless bees' species; (2) document the knowledge, management practices and local uses of products obtained from these insects; (3) the implementation of alternative management practices, and (4) the analysis of the availability of plant resources for stingless bees.

At the moment we identified a total of nine stingless bees' species in the study area [1]. Our inventory included a new registry of *Plebeia fulvopilosa*, Ayala (1999) for the region [2]. Local people use all the taxa identified since they provide honey, beeswax, and pollen. In a special way, *Melipona fasciata* Latreille (1811) has the highest local preference for the quality of its honey, while *Scaptotrigona hellwegeri* Friese (1900) and *Geotrigona acapulconis* Strand (1919) have a greater preference for their wax, since according to local people this species produces the highest amount of beeswax [3].

Stingless bees have been of great economic importance for the production of honey, a product that has been used since ancient times, and particularly in the case of Mexico, have been part of the social and religious life of different cultural groups [4].

Through our experience we have observed two relevant phenomena in the context of the study region, while is true that local knowledge about use, management and ecological issues on stingless bees is persistent, is being lost intergenerationally together with the disuse of the products of these bees, accompanied by the decline in stingless bee populations [3, 5]. On the other hand, new expressions are arising which re-signify stingless bees and their ecological, economic and social importance expressing in the emergence of local initiatives that seek not only to promote the production of stingless bees' honey, but also perspectives for their conservation [1].

In this context, we have oriented efforts toward the strengthening of the productive activity focusing on the honey characteristics that is being produced in the region. Primarily, we are interested in knowing the botanical origin of honey since its properties depend on the origin of the nectar or the secretions used by bees [6, 7]. Nectar (flower honey) are classified according to the main source where the bees collect the nectar being monofloral or multifloral [8, 9]. In addition, we consider relevant this source of information to understand the implications of environmental heterogeneity on the availability of resources for bees and their foraging patterns.

In a complementary way, we have been analyzing the physicochemical properties of honey. Recently the demand for natural, organic and homeopathic products has increased, including those of stingless bees. Although there are several works that seek to characterize the physicochemical properties particularly for the stingless bees' honey [10, 11] have not yet been defined formal norms or quality standards, which is very complex because of the variety of species and types of honey specific to each region. For this reason, it is important to deepen the understanding of the physicochemical stingless bees' honey characteristics.

From these perspectives, the aim of this chapter is to present the advances in the characterization of botanical origin of stingless bees' honey, and the analysis of their physicochemical properties in the study area.

2. Study area

The Balsas River Basin covers eight states, particularly for Michoacán this basin can be divided into three sub regions, taking an altitudinal criterion: Alto Balsas, Medio Balsas, and Bajo Balsas (Tepalcatepec). The specific area of study falls inside the sub-region of Alto y Medio Balsas. The process of collecting honey samples was carried out mainly in the municipality of Madero (19° 10′, and 19° 33′ N and 100° 59′, 101° 22′ W), it borders to the north with the municipalities of Acuitzio, Morelia and Tzitzio; to the east with Tzitzio and Tiquicheo de Nicolás Romero; to the south with Carácuaro and Nocupétaro; to the west with Nocupétaro, Tacámbaro, and Acuitzio. It has an altitudinal gradient that goes from 800 to 2900 m and has a temperate humid climate with summer rains (Cw) and in the south it is warm subhumid with summer rains (Aw). It has an annual rainfall of 1654.5 mm and its temperature ranges from 7.5 to 23.9°C [12]. In the municipality, the pine-oak forests are predominant and it presents a small portion of tropical dry forest. At the moment, a total of 62 species of api-botanical interest have been recorded, of which 65% provide nectar, 22% nectar and pollen, and 13% pollen to bees [13]. The most representative families are Asteraceae, Leguminosae, and Lamiaceae. It has been described that there are two flowering peaks, one that goes from March to May corresponding to the dry season and another that goes from September to November, which corresponds to the rainy season, specifically in the fall [13].

3. Collection and analysis of pot-honey and pot-pollen samples

For the analysis of the botanical origin, samples of pot-honey and pot-pollen were collected during May 2017 (dry season) and December 2017, from four species of stingless bees: *Melipona fasciata* (H-4691; H-4690), *Nannotrigona perilampiodes* (H-4686; H-4687), *Plebeia fulvopilosa* (H-4735; H-4736) and *Scaptotrigona hellwegeri* (H-4741; H-4742). Honey samples were obtained from hives under meliponiculture management in three communities of Michoacán State: Piumo, San Pedro Piedras Gordas and Etucuaro **Table 1**. The samples were cataloged and entered into the collection of the Palynology Laboratory of the Geology Institute of the National Autonomous University of Mexico, in which the melissopalynological process was performed. In general, the samples were processed by conventional chemical methods and to know the representativeness of each taxon the percentages of each pollen type were calculated from the count of 500 pollen grains per sample, in random transects [14]. The description and identification of the pollen grains was made under the microscope. The preliminary identification of pollen grains was carried out by comparison with the help of specialized keys from the Reference Palynological Collection of the Geology Institute. Honey was characterized as "monofloral" when in its composition a botanical species with pollen percentage ≥45% and "multifloral," mixed or polifloral prevailed when two or more species presented with percentages ≥10% [14].

We analyzed also honey physicochemical properties of the most frequently used stingless bees' species: *Melipona fasciata*, *Scaptotrigona hellwegeri*, *Geotrigona acapulconis*, *Frieseomelitta nigra*, *Plebeia fulvopilosa* and *Nannotrigona perilampoides* (of wild hives and hives under meliponiculture management). The physicochemical methods used to obtain the results of **Table 2** have been previously described [15]. Quantification of HMF and fructose was carried out with differential pulse polarography (DPP) using the calibration curve method. DPP is direct, and, because no other chemical reagents are used, lowers costs and health risks. On the other hand, the potentiometric monitoring of the progression of the Fehling reaction confirmed the reaction stoichiometry and produced more accurate and reproducible results. A principal component analysis (PCA) was used to identify relationships and patterns between honey samples and physicochemical characteristics.

4. Pot-pollen and pot-honey characteristics

4.1 Botanical origin based on melissopalynological analysis

Melissopalynological study showed polylectic behavior of *M. fasciata*, *N. perilampoides*, *P. fulvopilosa* and *S. hellwegeri* to collect pollen and nectar resources. Two monofloral honey samples of *Lopezia* sp. (52.5%) and *Paullinia* sp. (78.7%) were recorded for *P. fulvopilosa* and *S. hellwegeri* respectively. Moreover, honey sample of *M. fasciata* was bifloral of both Rhamnaceae (48%) and *Quercus* sp.

	Melipona fasciata		*Nannotrigona perilampiodes*		*Plebeia fulvopilosa*		*Scaptotrigona hellwegeri*	
	Honey	**Pollen**	**Honey**	**Pollen**	**Honey**	**Pollen**	**Honey**	**Pollen**
Date of collect	May/2017	May/2017	May/2017	May/2017	December/2017	December/2017	December/2017	December/2017
Localities	Piumo		San Pedro Piedras Gordas		Etucuaro		Etucuaro	
Num. catalog	(H-4691)	(H-4690)	(H-4686)	(H-4687)	(H-4735)	(H-4736)	(H-4741)	(H-4742)
Attalea sp.						66.9		
Alnus sp.			8.4					
Anacardiaceae			1.7					
Asteraceae			22.5	2.1	4.5			
Betulaceae	2.6							
Brassica sp.			3.0					
Bursera sp.			8.4					
Cyperaceae	2.2							
Fabaceae					12.3			
Fraxinus sp.	1.5	2.7	25.8		1.1			
Lopezia sp.					52.5			
Melastomataceae		10.1						
Paullinia sp.							78.7	
Polygalaceae					3.4			
Quercus sp.	38.5		18.8	96.6				
Rhamnaceae	48.0	85.3						
Rubiaceae						24.0		
Salix sp.							9.7	98.5

	Melipona fasciata		*Nannotrigona perilampiodes*		*Plebeia fulvopilosa*		*Scaptotrigona hellwegeri*	
	Honey	**Pollen**	Honey	Pollen	**Honey**	**Pollen**	Honey	Pollen
Date of collect	**May/2017**	**May/2017**	May/2017	May/2017	**December/2017**	**December/2017**	December/2017	December/2017
Localities	**Piumo**		San Pedro Piedras Gordas		**Etucuaro**		Etucuaro	
Num. catalog	**(H-4691)**	**(H-4690)**	(H-4686)	(H-4687)	**(H-4735)**	**(H-4736)**	(H-4741)	(H-4742)
Sapindaceae			1.5					
Sicyos sp.					3.4			
Solanum sp.	2.6							
Vernonia sp.	1.8							
Others	2.8	1.9	9.9	[illegible]	[illegible]	9.1	11.6	1.5

Table 1.
Percentages of botanical taxa recovered from pot-honey and pot-pollen samples in the melissopalynological analyses.

(38.5%). Because *Quercus* sp. produces honeydew, this honey (H-4691) might be considered that was made with a mixture of nectar of flowers and honeydews. On the other hand, pot-honey sample of *N. perilampoides* was multifloral, this bee species also combined nectar of Asteraceae (22.5%) and *Fraxinus* sp. (25.8%) flowers, as well as honeydews of *Quercus* sp. (18.8%). It is important to mention that *Quercus* sp. is also considered an excellent polliniferous tree (**Table 1**).

Respect to pot-pollen samples, the pot-pollen analyses gave evidence of polliniferous preferences of only one resource in all four stingless bee species. The best polliniferous plants were Rhamnaceae (85.3%) for *M. fasciata*; *Quercus* sp. (96.6%) was exploited by *N. perilampoides*; pollen from *Attalea* sp. (66.9%) was preferred by *P. fulvopilosa* and *Salix* sp. (98.5%) was chosen by *S. hellwegeri* (**Table 1**). The results also showed alternative resources, for instance, Anacardiaceae, *Brassica* sp., Melastomataceae, and Sapindaceae among others.

4.2 Physicochemical characteristics

In relation to the chemical parameters obtained from the stingless bees' honey samples analyzed (**Table 2**), higher moisture content was observed than the levels set in the Mexican standard (or practical values) for *A. mellifera* honey (≤20.0% of humidity). The registered pH is not >4 in the analyzed samples except for *N. perilampoides*. Also, found that the samples have a lower ash content than the value reported by the Mexican standard for *A. mellifera* honey (≤0.600% of ash content). The reducing sugars content is lower than the established by the Mexican standard for *A. mellifera* honey (≤63.88% of reducing sugars) because there was high levels of humidity, except in the case of the honey sample of *M. fasciata* which showed a higher value (65.73%) than *A. mellifera*. There were low levels of fructose and saccharose in the samples analyzed with respect to the parameters established for *A. mellifera* (≤45.0% of fructose and ≤8.0% of saccharose). For the HMF parameter was found that the samples tested are below of the maximum limits established in the Mexican standard for *A. mellifera* honey (≤80.0 mg kg^{-1}), and they are even below the detection limit, which reflects their high quality and their adequate conservation and storage process.

Just as there are differences between pot-honey and the honey of *A. mellifera*, at the same time a great variation in the characteristics of the different samples of pot-honey is observed. To exemplify, the moisture percentage ranges vary from a minimum value of 25.80% (*M. fasciata*) to a maximum value of 42.99% (*S. hellwegeri*) or the variation that occurs in reducing sugars that goes from 34.50% at its minimum value (*G. acapulconis*) and 65.73% as the maximum value (*M. fasciata*) (**Table 2**).

However, it is possible to observe certain relevant patterns. As shown in **Table 3**, the first two components of the PCA explain 66% of the variation in the data. The most important variables in the first component are the percentage of ash and pH, with antagonistic effects. In the second component, reducing sugars and pH are the most relevant variables, also with antagonistic interactions. In the biplot (**Figure 1**), two groups are distinguished; the pot-honey samples that were obtained from hives managed by meliponiculture techniques (M) are separated from the pot-honey samples that were obtained by direct extraction of wild nests (E).

There is a group of three samples of honey (M) that are strongly influenced by the reducing sugars and that showed the highest values of this parameter (*M. fasciata* and *S. hellwegeri*), while there is another group of three samples of honey (E) strongly influenced by HMF and that showed the highest values of this parameter (all TDF samples) (**Figure 1**).

Species	Mt	V	Se	Humidity (%)	pH	Ash content (%)	Reducing sugars (%)	F (%)	S (%)	HMF (mg kg^{-1})
A. mellifera	n/d	n/d	n/d	20.00	4.5	0.600	63.88	45.0	8.0	80.0
M. fasciata[1]	E	TDF	R	26.50	2.9	0.010	49.00	25.6	2.1	18.8
S. hellwegeri[1]	E	TDF	R	33.00	3.2	0.010	55.50	23.1	0.3	13.4
G. acapulconis[1]	E	TDF	D	35.60	3.2	0.240	34.50	31.1	1.4	9.6
F. nigra[1]	E	TDF	R	34.00	3.6	0.010	45.70	19.1	0.6	27.6
M. fasciata[2]	M	TMF	D	26.99	3.3	0.018	62.04	11.5	0.6	LDL
M. fasciata[2]	M	TMF	R	25.80	3.9	0.007	65.73	21.9	0.3	LDL
P. fulvopilosa[2]	M	TMF	D	35.40	3.8	0.348	47.89	22.1	0.4	LDL
P. fulvopilosa[2]	M	TMF	R	31.40	3.4	0.130	49.99	28.2	2.1	LDL
S. hellwegeri[2]	M	TMF	D	42.99	3.3	0.095	48.84	21.4	0.0	LDL
S. hellwegeri[2]	M	TMF	R	27.80	3.2	0.067	59.47	17.9	0.0	LDL
N. perilampoides[2]	M	TMF	D	25.00	4.3	0.251	42.93	15.2	0.0	LDL
N. perilampoides[2]	M	TMF	R	23.00	3.9	0.199	47.00	16.0	0.0	LDL

n/d, no data; Mt, management technique; E, extraction; M, meliponiculture; V, vegetation; TDF, tropical dry forest; TMF, transitional mixed forest; Se, season; D, dry season; R, rainy season; F, fructose; S, saccharose; LDL, lower than the detection limit. Values of A. mellifera according to the Mexican norm (NMX-F-036-1981, 2006); pH and fructose values for A. mellifera based on the maximum values reported by Bogdanov [16].
[1]The samples of honey were collected in 2011 in the municipality of Nocupétaro.
[2]The samples were collected in 2014 in the municipality of Madero.

Table 2.
Physico-chemical parameters of honey of six species of stingless bees, in the Alto Balsas, Michoacán, México.

	Comp.1	Comp.2	Comp.3	Comp.4	Comp.5	Comp.6	Comp.7
% of variance	36.78	29.18	13.69	12.31	4.62	2.13	1.29
% of cumulative variance	36.78	65.97	79.65	91.96	96.58	98.71	100.00
Loadings:							
Hum	0.258	0.330	0.725	0.277	0.132	0.449	0.000
pH	−0.429	0.339	−0.194	−0.348	0.639	0.352	−0.104
Ash	0.656	−0.218	−0.228	−0.151	0.662	0.000	0.000
RS	−0.304	−0.520	0.402	0.343	0.591	0.000	0.000
F	0.522	0.144	−0.183	0.284	0.614	−0.458	0.000
S	0.490	−0.105	−0.548	0.654	0.000	0.000	0.000
HMF	0.374	−0.210	0.233	−0.741	0.141	0.432	0.000

Hum, humidity; RS, reducing sugars; F, fructose; S, saccharose.

Table 3.
Importance of principal components and the relative contribution of original variables to PCA.

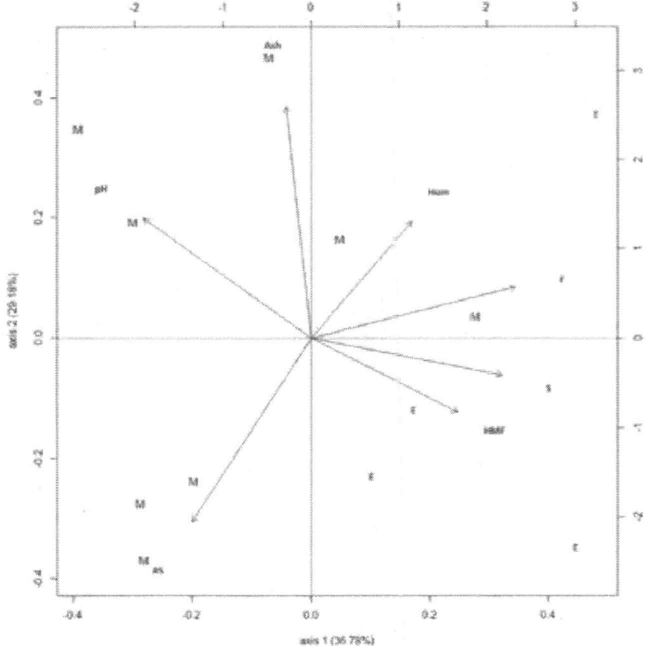

Figure 1.
Biplot for pot-honey samples.

5. Conclusions

In general terms, stingless bees from the study region use both arboreal and herbaceous species as resources, either from local vegetation or from introduced vegetation. It is interesting to note that *Quercus* sp. and *Fraxinus* sp. are not plants that require biotic pollination but nevertheless are a good source of food resources for bees. However, it is important to highlight the preliminary nature of the results obtained regarding the botanical origin of honey, this let us the goal of determining, at the level of species, the taxa presented in the melissopalynological samples of stingless bees' honey.

There is few information about physicochemical composition of stingless bee honey, and non-official quality control standards have been developed in Mexico or any country. Existing honey standards, for *A. mellifera*, in Mexico and the world cannot be applied to stingless bees' honey. In our samples, physicochemical parameters vary a lot, so which is very difficult to establish a norm. Differences in between species found in the study region is an indicator of the maturation and elaboration patterns of honey of each stingless bees species. This aspect should be considered in any attempt to promote a standardization of parameters of quality.

In a relevant way we documented in the pot-honey samples analyzed that they present an HMF content below the Mexican norm, and even below the limit of detection, which indicates that this honey is of excellent quality and had a good handling both when storing and when transporting. Regarding the moisture content, the analyzed samples presented high values not only with respect to the Mexican norm but also to other samples from other regions of Mexico. This may be due to the fact that honey was freshly collected with little storage time.

The content of total reducing sugars and fructose is higher in the honey of *A. mellifera*, which explains its greater commercialization. In the case of the pot-honey analyzed, it depends on the specie and in an important way, the vegetation type where stingless bees obtain their resources.

We consider that there is a need to generate information that allows us to establish general guidelines of quality for stingless bees' honey, and for this, it requires a wide sample of honey samples systematically generated including: diversity of species, ecosystems or vegetation types, seasonality of production and strategies of management to develop more precise statistical analyzes.

Acknowledgements

The work presented was done thanks to the financing of the "Programa de Apoyo a Proyectos de Investigación e Innovación Tecnológica (PAPIIT)" of the "Dirección General de Apoyo al Personal Académico (DGAPA)," through the project "Efectos del manejo y transformación del paisaje sobre las poblaciones de abejas nativas sin aguijón en la Cuenca del Balsas, Michoacán, México (TA200416)," National Autonomous University of Mexico. We would like to sincerely thank to the Meliponicultores Michoacanos del Balsas group and all meliponicultores and colmeneros of the study area for their help, interest and involvement

References

[1] Reyes-González A, Camou-Guerrero A, Gómez-Arreola S. From extraction to meliponiculture: A case study of the management of stingless bees in the West-Central Region of Mexico. In: Chambo D, editor. Beekeeping and Bee Conservation: Advances in Research. London: INTECH; 2016. pp. 201-223

[2] Reyes-González A, Ayala R, Camou-Guerrero A. Nuevo registro de abeja sin agujón del género Plebeia (Apidae: Meliponini), en el Alto Balsas del estado de Michoacán, México. Spain: Revista Mexicana de Biodiversidad. 2017;**88**:464-466

[3] Reyes-González A, Camou-Guerrero A, Reyes-Salas O, Argueta A, Casas A. Diversity, local knowledge and use of stingless bees (Apidae: Meliponini) in the municipality of Nocupétaro, Michoacan, Mexico. Journal of Ethnobiology and Ethnomedicine. 2014;**10**:47. DOI: 10.1186/1746-4269-10-47

[4] Quezada-Euán JJG, Nates-Parra G, Maués MM, Roubik DW, Imperatriz-Fonseca VL. The economic and cultural values of stingless bees (Hymenoptera: Meliponini) among ethnic groups of tropical America. Sociobiology. 2018;**65**(4):534-557

[5] Torres-Juárez AL. Saberes de niñas y niños sobre su territorio y abejas nativas sin aguijón de San Pedro Piedras Gordas, Madero, Michoacán [thesis]. Universidad Nacional Autónoma de México; 2018

[6] Norma mexicana, NMX-F-036-MORMEX-2006. Alimentos-miel-especificaciones y métodos de prueba. S.C., México, D.F: Sociedad Mexicana de Normalización y Certificación; 2006

[7] Tellería M, Forcone A. El polen de las mieles del Valle de Río Negro, Provincia Fitogeográfica del Monte (Argentina). Darwiniana. 2000;**38**(3-4):273-277

[8] Saínz-Laín C, Gómez-Ferreras C. Mieles españolas. Características e identificación mediante el análisis de polen. Editorial Mundi Prensa; 2000. 105 p

[9] Díaz-Forestier J, Gómez M, Montenegro G. Secreción de néctar de quillay. Una herramienta para una apicultura sustentable. Agronomía y Forestal. 2008;**35**:27-29

[10] Vit P, Medina M, Enríquez ME. Quality standards for medicinal uses of Meliponinae honey in Guatemala, Mexico and Venezuela. Bee World. 2004;**85**(1):2-5

[11] Souza BA, Roubik DW, Barth OM, Heard TA, Enríquez E, Carvalho C, et al. Composition of stingless bee honey: Setting quality standards. Interciencia. 2006;**31**(12):867-875

[12] Prontuario de información geográfica municipal de los Estados Unidos Mexicanos 2009. 2019. Available from: http://www3.inegi.org.mx/contenidos/app/mexicocifras/datos_geograficos/16/16049.pdf [Accessed: 30 April 2019]

[13] Romero Martínez DL. Disponibilidad temporal de la flora api-botánica en distintos tipos de vegetación, en el municipio de Madero, Michoacán, México [thesis]. Universidad Michoacana de San Nicolás de Hidalgo; 2017

[14] Córdova-Córdova CI, Ramírez-Arriaga E, Martínez-Hernández E, Zaldívar-Cruz JM. Caracterización botánica de miel de abeja (*Apis mellifera* L.) de cuatro regiones del estado de Tabasco, México, mediante técnicas melisopalinilógicas. Universidad y Ciencia. 2013;**29**(1):163-178

[15] Reyes-Salas EO, Gazcón-Orta ME, Manzanilla-Cano J, Reyes-Salas M, Camou-Guerrero A, Reyes-González A, et al. Electrochemical evaluation of quality characteristics in honey form Meliponini and *Apis mellifera* bees. Annals. Food Science and Technology. 2014;**15**(1):35-40

[16] Bogdanov S, Martin P, Lüllmann C. Harmonised methods of the European honey commission. Apidologie. Extra Issue. 1997:1-59

Chapter 3

Therapeutic Properties of Honey

Abstract

Honey has been used traditionally for ages to treat infectious diseases. These amazing properties of honey are complex as a result of the involvement of various bioactive compounds. Honey is becoming sustainable as a reputable and effective therapeutic agent to practitioners of conventional medicine and to the general public. Its beneficial role has been endorsed due to its antimicrobial, antiviral, anti-inflammatory, and antioxidant activities as well as boosting of the immune system. Also, other medical conditions discussed here which can be treated with honey include but not limited to diarrhea, gastric ulcer, canine recurrent dermatitis, diabetics, tumor, and arthritis, and honey can also be used for skin disinfection and wound healing. Most of the known factors that give honey these properties include its acidity, high sugar, hydrogen peroxide, and other non-peroxide properties. Some factors may affect the therapeutic properties of honey such as exposure to heat and light.

Keywords: antimicrobial, antiviral, wound healing, immune booster, skin infection, gastric ulcer

1. Introduction

Conventional medications have been utilized to treat infectious diseases for centuries, and one of the oldest remedies for microbial infection is honey. It has not been long that researchers rediscovered natural antimicrobial properties of honey [1]. Resistance to antibiotics is on the increase every day, and few new remedies are on the horizon, which led to further increased interest in the antimicrobial potency of honey. Many reports have shown that honey has antimicrobial activity against microorganisms such as protozoa, fungi, and bacteria, including viruses ([2], other references ought to be included). Despite the fact that bee honey is produced all around the world, its therapeutic properties may vary and are basically dependent on their entomological source (the type of bee), geographical location, and botanical origin (sources of nectars). Other external factors that may play some roles include but not limited to harvesting season, processing, storage condition, and environmental factors [1, 3]. The therapeutic potential of honey is greatly complex as a result of the action of various compounds as well as due to large variations in the concentrations of these compounds among honeys. The major biological properties that make it perfect as a therapeutic agent are antimicrobial (bactericidal or fungicidal), bacteriostatic (or fungistatic), anti-inflammatory potential, wound (sunburn healing) potential, antioxidant potential, radical scavenging activity, and antiviral activity [4–6]. Apart from boosting of the immune system, it can be used to treat other medical conditions such as diarrhea, gastric ulcer, canine recurrent dermatitis, diabetics, tumor, and arthritis and can also be used for skin disinfection and wound healing.

Honey is considered among the possible alternatives, which is natural, nontoxic, and with broad spectrum of action. This could be a promising substitute or supplement to antimicrobial agents, but some factors limit its use. Clinical applicability of honey has been hindered by incomplete knowledge of the antimicrobial activity and lack of precise mechanisms for determining the type of activity of honey, variations of honey, and its cost in some countries [2, 7].

2. Medicinal properties of honey

Proof from Stone Age paintings indicates that treatment of illnesses with honeybee began for over 8000 years ago. The use of honey as a medicine has been delineated by many historical records such as antiquated parchments, tablets, and books—Sumerian clay tablets (6200 BC), Veda (Hindu sacred text) 5000 years, Holy Bible, Koran, and Hippocrates (460–357 BC), and Egyptian papyri (1900–1250 BC) [7, 8]. The Qur'an clearly demonstrated the potential therapeutic value of honey. The Lord has roused the honeybees, to fabricate their hives on trees, in hills, and in man's residences; from inside their bodies comes a beverage of varying color, wherein there is recuperating for mankind, verily in this is a good signal, for the individuals who give thought [9, 10]. In spite of the fact that various articles have been published concerning honey, the vast majority of them have concentrated on the biochemical investigation, sustenance, and non-food business use. Honey was utilized for the treatment of many illnesses or disease conditions including asthma, eye diseases, tuberculosis, throat diseases, hiccups, unsteadiness, hepatitis, exhaustion, obstruction, thirst, piles, wounds, skin inflammation, worm invasion, and recuperation of ulcers (**Figure 1**) [2, 5, 8]. These properties are possible due to some of these potentials of honey to be discussed.

2.1 Antimicrobial activity

Therapeutically, the importance of antimicrobial activity of honey cannot be overemphasized, particularly in circumstance where the body's immune responses may be inadequate to clear disease or infection. In other words, honey has proven

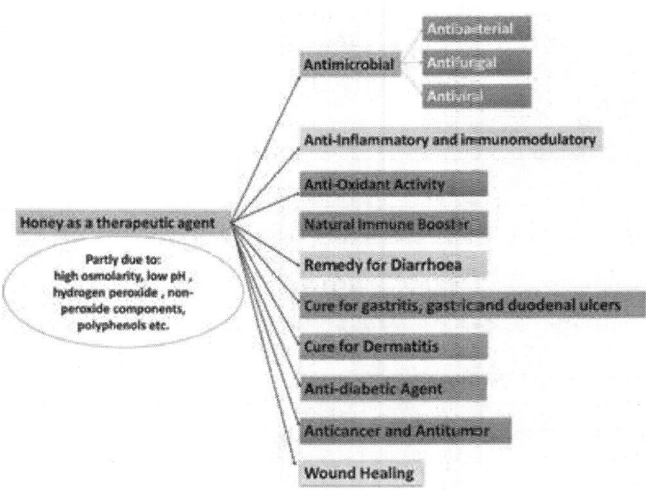

Figure 1.
Schematic representation of therapeutic potentials of bee honey.

to be an effective antimicrobial activity against both pathogenic and nonpathogenic microorganisms (such as bacteria, yeasts, and fungi), even in opposition to those microorganisms which have developed resistance to many antibiotics. The honey's antimicrobial effect could be bacteriostatic or bactericidal, relying upon the concentration used [4, 11]. Notwithstanding, its potentials have been credited to specific variables like high osmolarity (low water action), low pH (acidity), hydrogen peroxide (H_2O_2), and non-peroxide components [12, 13].

Moreover, bee honey is a solution of supersaturated sugar; these sugars prevent the thriving of microorganisms (bacteria and yeast) due to their high affinity for water molecules, thereby leaving little or no water to support their growth. As a result microbes become dehydrated and die in the end [11]. Naturally, the acidity of honey according to Fahim et al. prevents microbial growth, and usual pH of most of the pathogenic microbes ranges between 4.0 and 4.5 [14]. Be that as it may, the major antimicrobial potential has been reported to be due to hydrogen peroxide activity, a product of the glucose-oxidase enzyme oxidation of glucose, especially in diluted form of honey. The decomposition of hydrogen peroxide produces profoundly reactive free radicals, which respond and kill microbes. By and large, this honey property could easily be terminated in the presence of heat or due to catalase activity [15].

In any case, the antibacterial activity of some honeys may not always or necessarily be as a result of peroxide effect, but due to non-peroxide activity which results in a considerably more steady and stable antibacterial action. They are anyway called "non-peroxide honeys." Some examples of honey with non-peroxide activity are honeys from Australia (jelly bush—*Leptospermum polygalifolium*) and New Zealand (manuka honey—*Leptospermum scoparium*), which are hypothesized to have unidentified active component apart from the production of hydrogen peroxide. Unlike other honeys, they retain their potency to inhibit microbes even when catalase is present [2, 3].

It has been proposed that the principle part of this honeybee activity is probably of honeybee origin and partly due to plant origin. An appropriate solvent such as organic solvents (e.g., n-hexane, chloroform, ethyl acetate, and diethyl ether—by liquid-liquid or solid-phase extraction methods) could be used for extraction of the compounds exhibiting this activity. The separated mixes have been accounted for to incorporate flavonoids, unpredictable mixes (ascorbic corrosive, unbiased lipids, natural acids, carotenoid-like substances, and Maillard response items), phenolic acids, amino acids, and proteins [16, 17].

Other crucial effects of honey were related to its oligosaccharides. They have prebiotic properties, much like that of fructo-oligosaccharides. The oligosaccharides had been mentioned in reports to cause rise in population of some beneficial bacteria like bifidobacteria and lactobacilli, which are in charge of keeping up a sound intestinal microflora in human beings [18, 19]. In actuality, *Lactobacillus* spp. shield the body against some infections such as salmonellosis, and *Bifidobacterium* spp. limit the overgrowth of yeasts or bacterial pathogens in the gut wall, possibly lessening the danger of colon malignancy by out-contending putrefactive microbes equipped for freeing cancer-causing agents [18, 20].

The use of honey as a conventional remedy for microbial infections dates back to historical times [9]. There are reports on manuka (*L. scoparium)* honey, which has been proven to be effective against numerous human pathogens, inclusive of *S. aureus*, *Enterobacter aerogenes*, *Escherichia coli (E. coli)*, and *Salmonella* Typhimurium [13]. Some studies have found out that honey is very effective against methicillin-resistant *S. aureus* (MRSA), vancomycin-resistant enterococci (VRE), and streptococci [2]. Be that as it may, the recently identified bee honeys may have benefits over or similitudes with manuka honey because of improved antimicrobial potential local production (in this manner—accessibility) and more noteworthy selectivity against medically important organisms [12]. In terms of susceptibility to bee honey of comparable antibacterial efficiency, coagulase-negative staphylococci are very just like *S. aureus* [13] which according to Fahim et al. were more susceptible than *Pseudomonas aeruginosa* and *Enterococcus* species [14].

The disk diffusion method is for the most part a subjective test for identifying the vulnerability of microorganisms to antimicrobial substances; be that as it may, the minimum inhibitory concentration (MIC) mirrors the amount required for bacterial restraint. Following the in vitro strategies, many microscopic organisms (for the most part multidrug resistant; MDR) causing human diseases were discovered to be readily susceptible to honeys [3, 14, 16].

2.2 Anti-inflammatory and immunomodulatory activities

In spite of the fact that inflammation is a critical part of the regular response to infection or damaged tissues, when it is extreme or delayed, it can forestall healing or even cause further harm. The presently existing literature has shown that inflammatory reaction has been modulated in preliminary clinical studies, animal models, and cell cultures. The most serious outcome of immoderate inflammation is the production of free radicals within the tissue. These unfastened radicals are initiated by specific leucocytes that are stimulated as major aspect of the inflammatory process, as inflammatory processes are what activate the series of cellular events which precipitate to the initiation of growth factors that influence proliferation of fibroblasts, angiogenesis, and epithelial cells [21]. Several honey types from different countries have been reported to have anti-inflammatory effect, including honeys from stingless bees [4].

The anti-inflammatory effect of bee honey is due to its substantial amounts of phenolic contents. The repression of the pro-inflammatory actions of inducible nitric oxide synthase (iNOS) and/or cyclooxygenase-1 and cyclooxygenase-2 (COX-1 and COX-2) is caused by these phenolic and flavonoid compounds [22]. Moreover, when diluted natural bee honey is ingested, it results in the decrease of the prostaglandins' concentration including prostaglandin E_2 (PGE_2), thromboxane B_2 (in plasma of normal persons), and prostaglandin $F_2\dot{\alpha}$ (PGF2á) [6]. Strangely, in a colitis inflammatory model, honey became as effective as prednisolone remedy. While many adverse side effects of corticosteroids and NSAIDS, honey has natural anti-inflammatory effect free from major side effects [23].

Also, honey and its substances have been shown to be engaged with control of proteins, inclusive of iNOS, COX-2, tyrosine kinase, and ornithine decarboxylase [23, 24]. There are reports on the induction for the production of tumor necrosis factor alpha, interleukin-6 (IL-6), and IL-1β, by different types of honey [24, 25]. As of late, some honeys such as Gelam honey have been shown to reduce mediators of inflammatory reactions, for example, TNF-α and COX-2, by means of weakening NF-κB translocation to the nucleus and in this manner hindering the initiation of the NF-κB pathway. It is well known that the NF-κB activation performs a key function within the pathogenesis of inflammation. It is believed that production of fermentation agents such as short-chain fatty acid (SCFA) is a result of the slow absorption of honey, and SCFA has immunomodulatory activities, which have been proven to be so. It means that these fermentable sugars produced from honey such as nigerooligosaccharides have the ability to induce the immune response [11, 26]. Also, nonsugar ingredients present in honey may be responsible for immunomodulation [27].

Likewise, the application of honey topically has been found and reported in some published studies to lessen the quantity of exudate and edema in wounds, the two of which are identified with the action of wounds' local inflammatory process [28, 29]. All these and other studies imply that bee honeys have true anti-inflammatory and immunomodulation properties.

2.3 Antioxidant activity

In the human body, the bee honeys' antioxidant capacity is due to its ability to decrease oxidative reactions, which is estimated by its ability to scavenge free radicals [30]. It is believed that the anti-inflammatory action of honey could at least partly be due to its antioxidant activity since what is involved in various components of inflammation is oxygen free radicals [25]. Notwithstanding, when inflammatory process is not [31] directly stifled by honeys' antioxidant contents, they can be relied upon to scavenge free radicals so as to decrease the quantity of harm that would in any other case have resulted [32]. Honey is naturally composed of various flavonoids (including chrysin, pinocembrin, hesperetin, quercetin, apigenin, galangin, and kaempferol), Maillard reaction products and peptides, ascorbic acid, phenolic acids (such as ferulic, ellagic, caffeic, and p-coumaric acids), tocopherols, catalase, superoxide dismutase, and reduced glutathione, most of which provide a synergistic antioxidant effect by working together [33, 34].

The antioxidant action of honey is exerted by repressing free radical formation and usually catalyzed by some metal ions like copper, iron, etc. These metal ions in complexes can possibly be seized by some common constituents of honey such as flavonoids and other related polyphenols, thereby keeping the development of free radicals in the first place [25]. In terms of some sources of nutritional antioxidants, there are various phytochemicals in different honey varieties (just as other substances, for example, vitamins, organic acids, and enzymes) which may serve the purpose. The quantity and kind of those antioxidants depend mainly upon the sort of the honey and its floral source. All in all, it is now well known that darker honeys possess higher antioxidant content than lighter honeys [34, 35]. It has been shown that sugar analogue of about 14 unifloral honeys (which ranges from 3.0 to 17.0 μmol TE/g) had no antioxidant activity when examined using an assessment technique called oxygen radical absorbance capacity (ORAC). Reactive oxygen species (ROS) as well as free radicals are some of the contributing factors to some of the processes of disease and aging [31, 36].

Organisms shield themselves from those unfavorable compounds, to some extent, by retaining antioxidants from antioxidant-rich foods. In healthy human adults, this also depicts the impacts of taking 1.5 g/kg body weight of buckwheat/corn syrup honey on the antioxidant, including the reducing capacities of plasma. It very well may be estimated that these honey constituents could augment defenses against oxidative stress and that they may most likely shield us from oxidative pressure. Given that the normal sugar consumption by people is assessed to be more than 70 kg for each year, honey substitution in a few nourishments for traditional sugars could result in an upgraded antioxidant defense framework in healthful adults [33, 37, 38]. An Indian volatile oil of propolis (VOP) was researched using a photochemiluminescence strategy and spectrophotometric techniques, and it was discovered (from IC50 values) that the effectiveness of scavenging ABTS radicals by the VOP was increasingly articulated when contrasted with scavenging different radicals [39]. That is why many researchers in around the world have as well pressed for the consumption of the food highly-rich in antioxidants, such as honey [32].

It is critical to note that some factors such as botanical origin greatly affect the honeys' antioxidant activity; at the same time, its antioxidant capacity is only slightly affected by handling, processing, and storage condition of the honey. A strong correlation has been reported between the antioxidant activity and its total phenolic contents, including between antioxidant activity and the color of honey. The antioxidant activity according to many researchers may be located in both the water and ether fractions, which shows that the flavonoid contents of honey might be accessible to different compartments of the human body, wherein they may exert diverse physiological impacts [15, 36, 40].

2.4 Honey as a natural immune booster

Similarly, apart from honey having a direct antibacterial action, it could get rid of infection by immune system stimulation to fight the intruders. There is currently a sizeable report that honey is a natural immune booster. Many have reported that B60 lymphocytes and T lymphocytes can be activated to increase in number in cell culture and can as well activate neutrophils [15].

Moreover, Israili et al. in their investigation revealed that in cell cultures, monocytes can be stimulated to release the cytokines IL-1, IL-6, and TNF-alpha, the cell "messengers" which can activate numerous aspects of the immune reaction to infection [11]. Carter et al. in their review concluded that the production of TNF-α in macrophages by means of Toll-like receptor could be stimulated by a component of manuka honey (5.8 kDa) [2]. Jelly bush, manuka, and pasture honey, unlike artificial and honey-treated cells ($P < 0.001$), were reported to increase significantly the immune cells released from MM6 cells (and human monocytes) [22, 31]. Also according to a report on macrophages, honey is the main source of glucose that's essential for the "respiration burst" that results in hydrogen peroxide production, the dominant aspect of their bacteria-destroying activity [26, 29]. Besides, it is the major substrates for glycolysis, the principal mechanism for energy production in the macrophages, and hence lets them to function properly in broken tissue and exudates where the oxygen supply is often bad. The action of macrophages is also aided by the low pH of honey, which helps in the destruction of bacteria, so the acidity in phagocytic vacuole is what helps in digesting killed bacteria [15, 26]. A study revealed that intake of 80 g daily of natural honey for 21 days showed that in AIDS patients, prostaglandin levels were elevated compared with normal subjects. Finally, these studies are of the opinion that consumption of honey daily really improves one's immune system [41, 42].

2.5 Clinical conditions treatable with honey

2.5.1 As remedy for diarrhea

It is not unusual to say that intestinal tract infection occurs all throughout the world, affecting people of all ages. In diverse ways, dietary deficiencies are worsened by infectious diarrhea; however, as in any infection cases, there is an increase in calorific demand. A variety of microbes (bacteria, parasites, and viruses) may be responsible for an intense inflammation of the gastrointestinal tract, which leads to acute gastroenteritis [43]. Pure or unadulterated honey has shown to possess bactericidal action against numerous enteropathogenic microbes, such as enteropathogenic *E. coli*, *Shigella*, and *Salmonella* species [44, 45]. An in vitro study has shown that the mucosal epithelial cells' attachment by *Salmonella* species was prevented by honey; attachment is taken into consideration as the initial step in the establishment of gastrointestinal tract bacterial infection [46]. Apparently, the antibacterial activity of some honey varieties and its therapeutic usefulness against infections caused by *Salmonella* Typhimurium and *E. coli* 0157:H7 have been reported [11, 16].

Samarghandian et al. in their study reported that 30 mL of honey and a bland diet when administered three times a day were observed to be an effective cure in some patients (66%) and further relief was provided to 17% of them, while in over half of anemic patients, honey proved to be effective [47]. Gastroenteritis in infants and children was reported to have been with oral rehydration solution (ORS) and honey according to an investigation by Abdulrhman et al. [48]. In this study, there was a great reduction in the frequency of both bacterial and nonbacterial diarrhea. Most likely, it is actually less demanding, to add honey to ORS, which obviously made the solution a little bit sweet and perhaps increasingly adequate. Due to the fact that honey has high sugar content, it could be utilized for the promotion of water as well as sodium absorption from the bowel. When the intestinal mucosa is damaged, honey moreover assists in its repair, performs the function of an anti-inflammatory agent, and instigates the growth of new tissues [32, 48].

2.5.2 As medicine for gastritis and gastric and duodenal ulcers

Some of the health complications as a result of being infected by *Helicobacter pylori* are gastritis and gastric and duodenal ulcers. Traditional treatment for the annihilation of *H. pylori* is far from satisfactory; consequently there's search for alternative remedy. In the management of *H. pylori* infections, honey may contain a potential source of new compounds effective against the pathogen [6, 49]. Another in vitro study has shown that about 20% solution of honey was biocidal to gastritis-causing bacteria, *Helicobacter pylori*, isolated from a patient. Some of the isolates that were resistant to other antimicrobial agents were inhibited by honey solution [50, 51].

Furthermore, different honeys obtained from different countries and regions in an in vitro study by Ndip et al. using different honey concentrations showed that there were variations in the anti-*H. pylori* activity of the honeys [52]. This may be as a result of distinctive climatic conditions which may influence the distribution of vegetative species and flowers from which honeybees acquire nectar as well as sweet plant deposits for honey production. Because of hereditary heterogeneity displayed by *H. pylori* species, in aggregate with the regional variation in the antimicrobial components within the honey, there are differences in the honey concentrations that would be biocidal to *H. pylori* in specific locations. In particular, there is a report that *H. pylori* isolates from patients in Eastern Cape of South Africa were very susceptible to honey concentration as low as ≥10% v/v [53]. Apparently, it has been shown that honey dilution as low as 1:2 inhibited the isolates, but undiluted honey prominently inhibited the isolates [54]. Successful treatment of gastric ulcers with honey in the form of dietary supplement has been previously documented [6, 13]. Also, before the oral administration of ethanol, the use of honey orally or subcutaneously protected against gastric damage as well as reverses changes in pH induced by ethanol [32, 47].

2.5.3 As a medicine for dermatitis

Eczema, also known as atopic dermatitis (AD), is a common chronic atopic inflammation of the skin's outer layer. It is characterized by several skin problems, such as irritating skin, inflammation, blisters, redness, etc. Recurrent dermatitis occurs mostly in children, and it could as well be a problem in adults too; it makes an individual to be unattractive. Aside the one mentioned, there are other forms of dermatitis such as diaper dermatitis, seborrheic dermatitis, etc. In the pathogenesis of dermatitis generally, various immune cells participate such as macrophages, lymphocytes, eosinophils, and mast cells. Additionally, in the epidermis during this disease pathogenesis, keratinocytes play an important part due to its interaction with different cells of the immune system as well as stimuli from the external environment [55, 56].

One of the major problems in dermatitis patients (70–90%) and normal population (5%) is the skin colonization by *S. aureus*. This may be as a result of deficient cutaneous antimicrobial peptides, skin barrier function, as well as repeated scratching. Consequently, this bacterium is the major causative agent of superinfections of dermatitis lesions [57]. Also of importance to note are the productions of highly inflammatory substances such as exotoxins (α, β, γ, and δ cytolysins) and enterotoxins (SEA to SEE), which may play the role of superantigens as well as worsen the ongoing inflammation. Due to challenges in the management of the disease, natural remedies are opted for by many patients, and the topical use of honey on the lesions showed overall improvement in their symptoms [19, 55].

A study by Alangari et al. showed that manuka honey is effective in the treatment of dermatitis, particularly, atopic dermatitis, and after 7 days irrespective of honey treatment, there was no significant changes in the skin staphylococci. Also, they observed that honey in a dose-dependent manner downregulated IL4-induced CCL26 released from HaCaT cells significantly [58]. Another study has shown the effectiveness of honey mixed with olive oil and beeswax (in a ratio of 1:1:1 v/v) for the skin fungal infections, psoriasis, as well as dermatitis [59].

It has been reported that during the 7-day trial, honey has reduced the symptoms of diaper dermatitis and eradicated *C. albicans* from 50% of culture-positive patients. Probably, anti-inflammatory effects of the mixture could be the only reason for the symptomatic improvement in patients, due to the properties of the ingredients [60]. An investigation on *Staphylococcus aureus* isolated from patients with canine dermatitis revealed that honey had bactericidal effects against the test bacteria in vitro [61].

There is a postulate that in an in vitro study, natural raw honey, such as manuka honey, helps in the healing time through a dual effect on the inflammatory pathway. At first in the disease sites, honey is believed to suppress inflammatory cells' production and migration. Again, it allows normal healing process to occur through the epithelial cells and fibroblast proliferation enhancement and the production of pro-inflammatory cytokines [62]. Natural honey from different floral varieties possesses variable moisture content, depending on the quantity of water they contain (ranges between 6 and 14%), and provides the needed moisture to the inflamed skin without causing maceration [19, 62]. Nonetheless, more investigation is still needed such as randomized controlled trials, so as to find the most effective duration, frequency, and type of honey.

2.5.4 As antidiabetic agent

Diabetes mellitus is still a serious issue which is related with poor quality of life, cardiovascular intricacies, and increased mortality and morbidity. As a result of its economic and social burdens, it is becoming a public health concern. In humans, the most common forms of diabetes are type 1 and type 2 diabetes. The former results when the insulin is destroyed by the host immune system, while the latter which is most prevalent and genetically determined may be as a result of several factors. Despite the fact that diabetes has no known cause, complex interaction of a few factors such as environmental, social, and genetic components is involved in its etiology [63]. At the moment, the obtainable antidiabetic drugs are far from being satisfactory, due to some limitations such as the cost and availability. Alternatively, some patients have resorted to the use of dietary supplements or components, herbal preparations, and other natural and apicultural products such as honey [5].

Predominantly, honey is made up of monosaccharaides (fructose and glucose) and water as well as other components (more than 200). For quite a while, some people have the notion that diabetic patients cannot consume honey due to its high sugar content. This has given rise to a number of questions, such as "Can honey replace sugar in diabetic diet? Is sugar in honey essential in the prevention and treatment diabetes mellitus?." But many researchers around the globe have been working on honey characterization from different sources and the determination of its biological properties for a long period. Different studies have acknowledged the effectiveness of honey and its use in diabetes mellitus patients, including animal model studies, preclinical and clinical studies, and human studies [64].

In a natural honey, fructose content and the fructose/glucose ratio range from 21 to 43% and 0.4–1.5 (or even higher), respectively. The glycemic index of fructose, glucose, and sucrose (refined sugar) are, respectively, 19, 100, and 60, even though the naturally occurring sweetener and the sweetest is fructose. Although the mechanism of hypoglycemic effect of honey is still unknown, various studies have proven it to be so [8, 65]. In animal model experiments, reduction in blood glucose due to fructose has been reported, and it is believed that it may be as a result of reduced food intake, reduced rate of intestinal absorption, and prolongation of gastric emptying time. In hepatocytes, fructose activates glucokinase, which is needed in the assimilation as well as storage of glucose as glycogen by the liver. Glucose unlike fructose boosts the fructose absorption and aids in its hepatic actions through strengthening its delivery to the liver. The pancreas is an essential organ in diabetes, in that it produces insulin and glucagon; the antioxidant molecules in honey help in its protection against oxidative stress and damage, which may be another likely mechanism of hypoglycemic effect of honey. The insulin response and glucose homeostasis in normal rats are ameliorated due to intake of only fructose or in combination with sucrose molecule, compared to rats which received glucose. The hypoglycemic effect of honey has been demonstrated using various animal models, such as type 1 and type 2 diabetes induction in alloxan and streptozotocin using appropriate doses [64, 66, 67].

In another investigation, honey and fructose were used to feed diabetic (alloxan-induced) and healthy rats, respectively, and it was reported that the former had significant reduction in glucose level while it was not significant in the latter [68]. Honey proved its potency compared with sucrose and dextrose when included in the diet of diabetic (or hypertriglyceridemia) and healthy patients. There was a decrease in the elevated and normal C-reactive protein, homocysteine value, and triacylglycerol (in hypertriglyceridemia patients), and lipid profile was improved. The rise in plasma glucose level in diabetic patients was significantly reduced with bee honey compared with dextrose. Honey unlike sucrose made the insulin level to rise; in normal subjects, there was reduction in C-reactive protein, blood lipids, and homocysteine, after consumption of honey at different time [47, 69]. In summary, honey may be very effective in the management of patients with diabetes mellitus based on the evidence from experimental studies. Although very few reports have contrary view on the use of honey, most researchers believe that it is very useful in reducing metabolic disorders, managing hyperglycemic state, and reducing diabetic complications on different organs.

2.5.5 Anticancer and antitumor activity

The potential to induce genetic mutation is called mutagenicity, which is interlinked with carcinogenicity. The heterocyclic amines such as Trp-p-1 (3-amino-1, 4-dimethyl-5H-pyridol [4,3-b] indole) are formed especially during food frying and roasting processes. Many studies around the world have demonstrated the anticancer potential of honey in tissue cultures [70, 71], in animal models, and in clinical trials. One of the main active constituent of honey responsible for its anticancer activity is believed to be the polyphenols. Some of the anticancer properties of honey apart from its anti-inflammatory, antioxidant, and immunomodulatory activities are due to its antiapoptotic, antiproliferative, antitumor, antimutagenic, and estrogenic modulatory activities.

Programmed cell death and cellular proliferation (uncontrolled) are the main features of cancer cells. Fauzi et al. reported that through mitochondrial membrane depolarization, honey is able to induce apoptosis (programmed cell death) in various types of cancer cells [71]. In human colon cancer cell lines, the high tryptophan and phenolic content of honey induces programmed cell death by upregulating the expression of proapoptotic proteins (caspase 3, p53, and Bax) and modulating the expression of antiapoptotic proteins (Bcl-2) [72]. Manuka honey-induced programmed cell death involves activation of PARP, loss of Bcl-2 expression, and induction of DNA fragmentation [73].

Throughout human and animal life, there is division of epithelial cell, and in this cell cycle, G_1/S phase transition regulates cell growth. The loss of this regulation leads to tumor/cancer. Honey is very promising in arresting the cell cycle. A study has reported that honey supplemented with *Aloe vera* solution significantly reduced the expression of tumor cell proliferation (nuclear protein—Ki67-LI) in rat by arresting the cell cycle [74]. The antiproliferative activity of honey and its components (like flavonoids and phenolics) in G_0/G_1 phase cell cycle has been reported in colon [75], melanoma [76], and glioma [77] cancer cell lines. This potential of honey has been confirmed to be dose- and time-dependent [76].

The multifunctional signaling protein, tumor necrosis factor, plays vital beneficial and deleterious roles in diverse cellular events including initiation, promotion, and progression of tumor cell. In vitro and in vivo in mice, the antitumor effect of honey has shown that it is effective in inhibiting the growth of different bladder cancer cell lines (T24, RT4, 253 J, and MBT-2), and when administered orally or intralesionally in the bladder cancer (MBT-2) implantation mice models, a good result was obtained. Royal jelly honey proteins (apalbumin-1 and apalbumin-2) according to Šimúth et al. [78] stimulate macrophages to release cytokines interleukin-1 (IL-1), interleukin-6 (IL-6), as well as TNF-α. Different honeys at very low concentration (1% w/v) such as manuka, jelly bush, and pasture honey initiate release of the TNF-α and interleukin- (IL-) 1β and IL-6 [79, 80].

Researchers have shown that honey has a strong antimutagenic property and hence has anticarcinogenic potential [81, 82]. The cells of *E. coli* exposed to UV or γ radiation showed that honey elicited SOS response (SOS is an error-prone repair pathway contributing to mutagenicity). This study on genes involved in SOS-mediated mutagenesis including *umuC*, *recA*, and *umuD* demonstrated that honey inhibited the changes significantly, thus confirming the strong antimutagenic effect of honey [82]. Another study that used Ames assay to investigate the antimutagenic activity of honeys (acacia, buckwheat, Christmas berry, soybean, tupelo, and fireweed) against Trp-p-1 in comparison with sugar analogues reported that they exhibited a significant inhibition of Trp-p-1 mutagenicity [83]. The antitumor properties of honey and its possible mode antimetastatic action were investigated using anaplastic colon adenocarcinoma of Y59 rats and spontaneous mammary carcinoma in methylcholanthrene-induced fibrosarcoma of CBA mice. This study showed that oral administration of honey produced statistically significant antimetastatic effect. Their findings showed that apart from the immune cells' activation, ingestion of honey may be more beneficial with respect to cancer and metastasis prevention [84, 85]. Aside from all of these, more research such as randomized controlled clinical trials is still required to validate the uses of honey either alone or as adjuvant therapy for cancer and its associates.

2.5.6 The action of honey in wound healing

A wound is said to result when there is an interruption of the progression of a tissue structure. Wound healing, a continuous as well as complex process, has three stages, viz., inflammation, a proliferative phase, and tissue remodeling. It is essentially the aftereffect of interactions among blood, growth factors, cellular elements, cytokines, and the extracellular matrix [15, 86].

The healing properties of honey have long been recognized and documented [24]. Both endogenous (pathophysiology) and exogenous (microorganisms) factors affect the healing of wounds. Due to the local conditions of the wound environment, there is increased risk of infection of the wound by pathogens. Many bacterial species have been recouped from infected wounds; however, *Staphylococcus aureus* is the most as often as possible isolated. *Pseudomonas aeruginosa* is additionally an essential pathogen in chronic wounds and burns; this has been reported in various investigations and has been found in one third of chronic leg ulcers [5, 47]. The use of honey for treatment of infected wounds and normal wounds has been reported in many articles [5, 6]. Honey with demonstrated antibacterial action can possibly be a compelling treatment choice for wounds infected or at risk of infection with different human pathogens.

The use of honey in the treatment of wounds as a result of skin ulcers resulting from various etiologies has been documented in literatures [6]. A review by Jull et al. showed that out of over 470 cases observed in which honey was used as a therapeutic agent, successful healing was not achieved only in five cases [87]. Honey exerted both deodorizing effect and anti-inflammatory actions on the wound and thus reduced the level of pain. A study carried out in the UK which tried to determine the effect of medical honey on three patients suffering from chronic leg ulceration proved the honey treatment to be effective. Even though they all had some years of disease reoccurrence, there was significant healing in all cases with a decrease in occurrence rate, pain, and discomfort [88]. In another study by Dunford and Hanano, Medihoney dressings were used on the leg ulcers of 40 patients for a 12-week study period. These ulcers had previously been subjected to 40 patients whose leg ulcers had not responded to 12 weeks of compression therapy with no recorded improvement. However, after treatment with Medihoney, there was a remarkable decrease in the ulcer pain and size, and the odorous wounds were promptly deodorized [89]. Another study in Ireland done to qualitatively determine the bacteriological changes that occurred in a 4-week treatment period with either a hydrogel dressing or manuka honey enrolled 108 study subjects. Methicillin-resistant *S. aureus* was identified in 16 of the patients, 10 in the honey group and 6 in the hydrogel group. The authors reported that at the end of the 4-week period, manuka honey showed its potency in eradicating MRSA from 70% of patients with chronic venous ulcers [90].

Burn injuries are typically connected with a high occurrence of death and disability. Advances in biology of cells and knowledge in wound healing as well as growth factors have aided in the burn injuries' management. Split-thickness skin grafting with autografts is a well-known standard of care for burn wounds. Researches that investigated the use of honey in the treatment of burns have been documented [6, 37]. A review by Zbuchea made mention of a study carried out in France, in which there was rapid healing for first- and second-degree burns, and in Netherlands, honey-treated burns were found to show less inflammation than those treated with sugar and silver sulfadiazine [91]. In a randomized study in Pakistan, the efficacy of honey for the treatment of superficial and partial-thickness burns covering <40% of human body surface area was determined in 150 patients, and the results compared with those of silver sulfadiazine. The re-epithelialization rate and healing of both superficial and partial-thickness burns were remarkably faster in the sites treated with honey than in the sites treated with silver sulfadiazine. *P. aeruginosa* was isolated from 6 patients in the honey-treated site, while 27 patients had positive culture in the silver sulfadiazine-treated site [92].

The infection of the wound is a critical factor that delays or hinders wound healing. Honey has many properties, both antibacterial and otherwise, that enhance its beneficial effects on wound healing [6]. The properties of honey that give it its wound healing power have been previously discussed, such as hydrogen peroxide which is an important antiseptic and stimulant of wound healing process.

Other property is due to its high osmolarity which inhibits microbial growth in light of the fact that the sugar molecules tie up water molecules such that there is insufficient water for the microbes to grow. Research has shown that the application of antioxidants to burns reduces inflammation [38]. Honey inactivates the free iron, which is believed to catalyze the oxygen free radical formation produced as a result of H_2O_2, and its antioxidant components aid in the removal of oxygen free radicals [93]. Apart from various immune system cells' stimulations and nitric oxide end product in honey, honey provides a protective barrier and, by osmosis, set up a soggy wound recuperating condition that does not adhere to the underlying wound tissues.

3. Factors that may affect the therapeutic potentials of honey

The destruction of therapeutic properties of honey by exposure to light and heat was first reported by Dold and Witzenhausen, who found that inhibine in honey was not stable [94]. Other numerous reports have confirmed this finding, but variations in the properties of honey as a result of heat and light presently reported vary. On exposure of honey samples heated at 56°C for 30 min, 80°C for 10 min, and 100°C for 5 min, 17% of the samples partly lost their inhibitory activity against the test organisms [95]. Also, a complete loss of activity has been reported. The antimicrobial activity of honey, both new and stored, reduced after heating at 80°C for 1 h. Also, long storage of honey (5 years) reduced this activity [96]. In some instances, when honey is subjected to lesser degrees of heating, the activity is retained which may be due to partial destruction of the heat-sensitive factor [97]. Similarly, exposure of honey to heat at 46°C for 8 h, 52°C for 8 h, and 55°C for 8 h resulted in the increase in the MIC from 4 to 8%, to 12%, and to 16%, respectively [98]. Another investigation showed that when honeys are held at 40°C for 96 h and 37°C for 24 h, there was no reduction in its antibacterial activity, as in the cases previously mentioned. This is possible since the temperature in the beehive where honey can spend quite a long time is around 34°C. At the same condition, diluted honey may not be stable, since the hydrogen peroxide production rate drops off with time [96]. The variations in antimicrobial action of honey due to heat is believed to depend mainly on pH of the honey, and at low pH, activity can rapidly be lost [98].

The effect of light on honey therapeutic potentials has been reported long ago. This observation has been confirmed by many researchers [99, 100]. The non-osmotic activity of honey has been reportedly lost after its exposure in a layer 1–2 mm thick to sunlight for 15 min [100]. Honey stored on a window sill for 8 months in 2.5 L transparent polystyrene jars completely lost its activity, but when stored in white polyethylene jars for the same period, its activity remained the same [101]. This shows that protecting honey from sunlight or UV light above 400 nm will prolong its activity [94]. But Molen reported that there was no reduction of activity when a thin film of honey solution was exposed for 1 h to an ultraviolet (UV) lamp (254 nm) [102]. Unlike light-colored honey, the light stability of dark-colored honey has been reported, and it is believed to be as a result of reduction in light wave reaching the bulk of the honey [102].

4. Conclusions

For ages, honey has been traditionally used to treat human diseases. Recently, it is becoming acceptable to everybody as a therapeutic agent that is cost-effective and lacks side effects. The therapeutic and beneficial properties of honey have been endorsed due to:

- Its antibacterial, antifungal, antiviral, and antiparasitic activities against a wide range of organisms.
- Its anti-inflammatory effect and immunomodulatory activities due to its substantial amounts of phenolic contents
- Its antioxidant capacity, which is estimated by its ability to scavenge free radicals.
- Its natural immune-boosting capability.
- In addition to these properties honey could also be used to treat other medical conditions which include but not limited to:
- Diarrhea caused by enteropathogenic *E. coli*, *Shigella*, and *Salmonella* species.
- Gastritis and gastric and duodenal ulcers which may be as a result of *H. pylori* infections.
- Canine recurrent dermatitis, diaper dermatitis, and seborrheic dermatitis.
- Diabetics, where honey has been shown to be very effective in the management of patients with diabetes mellitus based on the evidence from animal model studies, preclinical and clinical studies, and human studies.
- Cancer and tumor—which may be due to its antiapoptotic, antiproliferative, antitumor, antimutagenic, and estrogenic modulatory activities.
- Wounds—its action on wound healing has been medically accepted especially in diabetic patients.

Most of the known factors that give honey these properties include its acidity, high sugar, hydrogen peroxide, and other non-peroxide properties. But some other factors may affect the therapeutic properties of honey such as:

- Its exposure to heat or higher temperature
- Its exposure to light, sunlight, or UV light

In all, honey has an interesting potential as a therapeutic agent which might in the near future complement/replace conventional drugs such as antibiotics. Therefore, more research is still needed to finally establish the basis for classifying honey as medical grade.

References

[1] Erler S, Moritz RFA. Pharmacophagy and pharmacophory: Mechanisms of self-medication and disease prevention in the honeybee colony (Apis mellifera). Apidologie. 2016;**47**(3):389-411

[2] Carter DA, Blair SE, Cokcetin NN, Bouzo D, Brooks P, Schothauer R, et al. Therapeutic manuka honey: No longer so alternative. Frontiers in Microbiology. 2016;7(569):1-11

[3] Oryan A, Alemzadeh E, Moshiri A. Biological properties and therapeutic activities of honey in wound healing: A narrative review and meta-analysis. Journal of Tissue Viability. 2016;**25**(2):98-118

[4] Nweze JA, Okafor JI, Nweze EI, Nweze JE. Comparison of antimicrobial potential of honey samples from Apis mellifera and two stingless bees from Nsukka, Nigeria. Journal of Pharmacognosy and Natural Products. 2016;**2**(4):1-7

[5] Alam F, Islam MA, Gan SH, Khalil MI. Honey: A potential therapeutic agent for managing diabetic wounds. Evidence-based Complementary and Alternative Medicine. 2014;**2014**:1-16

[6] Al-Waili N, Salom K, Al-Ghamdi AA. Honey for wound healing, ulcers, and burns; data supporting its use in clinical practice. Scientific World Journal. 2011;**11**:766-787

[7] Meo SA, Al-Asiri SA, Mahesar AL, Ansari MJ. Role of honey in modern medicine. Saudi Journal of Biological Sciences. 2017;**24**(5):975-978

[8] Abd Jalil MA, Kasmuri AR, Hadi H. Stingless bee honey, the natural wound healer: A review. Skin Pharmacology and Physiology. 2017;**30**(2):66-75

[9] Purbafrani A, Amirhosein S, Hashemi G, Bayyenat S, Moghaddam HT, Saeidi M. The benefits of honey in holy Quran. International Journal of Pediatrics. 2014;**2**(5):67-73

[10] Koshak A, Alfaleh A, Abdel-Sattar E, Koshak E. Medicinal plants in the holy Quran and their therapeutic benefits. Planta Medica. 2012;**78**(05):109

[11] Israili ZH. Antimicrobial properties of honey. American Journal of Therapeutics. 2014;**21**(4):304-323

[12] Alvarez-Suarez J, Gasparrini M, Forbes-Hernández T, Mazzoni L, Giampieri F. The composition and biological activity of honey: A focus on Manuka honey. Food. 2014;**3**(3):420-432

[13] Almasaudi SB, Al-Nahari AAM, Abd El-Ghany ESM, Barbour E, Al Muhayawi SM, Al-Jaouni S, et al. Antimicrobial effect of different types of honey on *Staphylococcus aureus*. Saudi Journal of Biological Sciences. 2017;**24**(6):1255-1261

[14] Fahim H, Dasti JI, Ali I, Ahmed S, Nadeem M. Physico-chemical analysis and antimicrobial potential of Apis dorsata, Apis mellifera and Ziziphus jujube honey samples from Pakistan. Asian Pacific Journal of Tropical Biomedicine. 2014;**4**(8):633-641

[15] Molan P, Rhodes T. Honey: A biologic wound dressing. Wounds-A Compendium of Clinical Research and Practice. 2015;**27**(6):141-151

[16] Julianti E, Rajah KK, Fidrianny I. Antibacterial activity of ethanolic extract of cinnamon bark, honey, and their combination effects against acne-causing bacteria. Scientia Pharmaceutica. 2017;**85**(2):19

[17] Kaur R, Kalia P, Kumar NR, Harjai K. Preliminary studies on different extracts of some honey bee products. Journal of Applied and Natural Science. 2013;**5**(2):420-422

[18] Cornara L, Biagi M, Xiao J, Burlando B. Therapeutic properties of bioactive compounds from different honeybee products. Frontiers in Pharmacology. 2017;**8**:412

[19] Burlando B, Cornara L. Honey in dermatology and skin care: A review. Journal of Cosmetic Dermatology. 2013;**12**(4):306-313

[20] Finegold SM, Li Z, Summanen PH, Downes J, Thames G, Corbett K, et al. Xylooligosaccharide increases bifidobacteria but not lactobacilli in human gut microbiota. Food & Function. 2014;**5**(3):436

[21] Aziz M, Jacob A, Yang W-L, Matsuda A, Wang P. Current trends in inflammatory and immunomodulatory mediators in sepsis. Journal of Leukocyte Biology. 2013;**93**(3):329-342

[22] Viuda-Martos M, Ruiz-Navajas Y, Fernández-López J, Pérez-Álvarez JA. Functional properties of honey, propolis, and royal jelly. Journal of Food Science. 2008;**73**(9):R117-R124

[23] Nooh HZ, Nour-Eldien NM. The dual anti-inflammatory and antioxidant activities of natural honey promote cell proliferation and neural regeneration in a rat model of colitis. Acta Histochemica. 2016;**118**(6):588-595

[24] Hadagali MD, Chua LS. The anti-inflammatory and wound healing properties of honey. European Food Research and Technology. 2014;**239**(6):1003-1014

[25] Yuksel P. Comparison of the VersaTrek and BACTEC MGIT 960 systems for the contamination rate, time of detection and recovery of mycobacteria from clinical specimens. African Journal of Microbiology Research. 2011;**5**(9):985-989

[26] Mesaik MA, Dastagir N, Uddin N, Rehman K, Azim MK. Characterization of immunomodulatory activities of honey glycoproteins and glycopeptides. Journal of Agricultural and Food Chemistry. 2015;**63**(1):177-184

[27] Alvarez-Suarez J, Giampieri F, Battino M. Honey as a source of dietary antioxidants: Structures, bioavailability and evidence of protective effects against human chronic diseases. Current Medicinal Chemistry. 2013;**20**(5):621-638

[28] Rao PV, Krishnan KT, Salleh N, Gan SH. Biological and therapeutic effects of honey produced by honey bees and stingless bees: A comparative review. Revista Brasileira de Farmacognosia. 2016;**26**(5):657-664

[29] Majtan J. Honey: An immunomodulator in wound healing. Wound Repair and Regeneration. 2014;**22**(2):187-192

[30] Osés SM, Pascual-Maté A, Fernández-Muiño MA, López-Díaz TM, Sancho MT. Bioactive properties of honey with propolis. Food Chemistry. 2016;**196**:1215-1223

[31] Anand S, Pang E, Livanos G, Mantri N. Characterization of physico-chemical properties and antioxidant capacities of bioactive honey produced from Australian grown Agastache rugosa and its correlation with colour and poly-phenol content. Molecules. 2018;**23**(1):108

[32] Nasuti C, Gabbianelli R, Falcioni G, Cantalamessa F. Antioxidative and gastroprotective activities of anti-inflammatory formulations derived from chestnut honey in rats. Nutrition Research. 2006;**26**(3):130-137

[33] Moniruzzaman M, Sulaiman S, Khalil M, Gan S. Evaluation of physicochemical and antioxidant properties of sourwood and other Malaysian honeys: A comparison with manuka honey. Chemistry Central Journal. 2013;**7**(1):138

[34] Chaikham P, Prangthip P. Alteration of antioxidative properties of longan flower-honey after high pressure, ultra-sonic and thermal processing. Food Bioscience. 2015;**10**:1-7

[35] Khalil MI, Sulaiman SA, Boukraa L. Antioxidant properties of honey and its role in preventing health disorder. The Open Nutraceuticals Journal. 2010;**3**(1):6-16

[36] Can Z, Yildiz O, Sahin H, Akyuz Turumtay E, Silici S, Kolayli S. An investigation of Turkish honeys: Their physico-chemical properties, antioxidant capacities and phenolic profiles. Food Chemistry. 2015;**180**:133-141

[37] Tahir HE, Xiaobo Z, Zhihua L, Jiyong S, Zhai X, Wang S, et al. Rapid prediction of phenolic compounds and antioxidant activity of Sudanese honey using Raman and Fourier transform infrared (FT-IR) spectroscopy. Food Chemistry. 2017;**226**:202-211

[38] Schramm DD, Karim M, Schrader HR, Holt RR, Cardetti M, Keen CL. Honey with high levels of antioxidants can provide protection to healthy human subjects. Journal of Agricultural and Food Chemistry. 2003;**51**(6):1732-1735

[39] Ramnath S, Venkataramegowda S. Antioxidant activity of Indian propolis—An in vitro evaluation. International Journal of Pharmacology, Phytochemistry and Ethnomedicine. 2016;**5**:79-85

[40] Kassim M, Achoui M, Mustafa MR, Mohd MA, Yusoff KM. Ellagic acid, phenolic acids, and flavonoids in Malaysian honey extracts demonstrate in vitro anti-inflammatory activity. Nutrition Research. 2010;**30**(9):650-659

[41] Al-Waili NS, Al-Waili TN, Al-Waili AN, Saloom KS. Influence of natural honey on biochemical and hematological variables in AIDS: A case study. Scientific World Journal. 2006;**6**:1985-1989

[42] Wan Yusuf WN, Wan Mohammad WMZ, Gan SH, Mustafa M, Abd Aziz CB, Sulaiman SA. Tualang honey ameliorates viral load, CD4 counts and improves quality of life in asymptomatic human immunodeficiency virus infected patients. Journal of Traditional and Complementary Medicine. 2018:1-8

[43] Graves NS. Acute gastroenteritis. Primary Care; Clinics in Office Practice. 2013;**40**(3):727-741

[44] Mohapatra DP, Thakur V, Brar SK. Antibacterial efficacy of raw and processed honey. Biotechnology Research International. 2011;**2011**:1-6

[45] Aggad H, Guemour D. Honey antibacterial activity. Medicinal and Aromatic Plants. 2014;**3**(2):152. DOI: 10.4172/2167-0412.1000152

[46] Alnaqdy A, Al-Jabri A, Al Mahrooqi Z, Nzeako B, Nsanze H. Inhibition effect of honey on the adherence of salmonella to intestinal epithelial cells in vitro. International Journal of Food Microbiology. 2005;**103**(3):347-351

[47] Samarghandian S, Farkhondeh T, Samini F. Honey and health: A review of recent clinical research. Pharmacognosy Research. 2017;**9**(2):121-127

[48] Abdulrhman MA, Mekawy MA, Awadalla MM, Mohamed AH. Bee honey added to the oral rehydration solution in treatment of gastroenteritis in infants and children. Journal of Medicinal Food. 2010;**13**(3):605-609

[49] Graham DY. History of helicobacter pylori, duodenal ulcer, gastric ulcer and gastric cancer. World Journal of Gastroenterology. 2014;**20**(18):5191

[50] Kim S, Hong I, Woo S, Jang H, Pak S, Han S. Isolation of abscisic acid from korean acacia honey with anti-helicobacter pylori activity. Pharmacognosy Magazine. 2017;**13**(50):170

[51] Nzeako BC, Al-Namaani F. The antibacterial activity of honey on helicobacter pylori. Sultan Qaboos University Medical Journal. 2006;**6**(2):71-76

[52] Ndip RN, Malange Takang AE, Echakachi CM, Malongue A, Akoachere J-FTK, Ndip LM, et al. In-vitro antimicrobial activity of selected honeys on clinical isolates of helicobacter pylori. African Health Sciences. 2007;**7**(4):228-232

[53] Ayala G. Exploring alternative treatments for helicobacter pylori infection. World Journal of Gastroenterology. 2014;**20**(6):1450

[54] Manyi-Loh CE, Clarke AM, Munzhelele T, Green E, Mkwetshana NF, Ndip RN. Selected South African honeys and their extracts possess in vitro anti-Helicobacter pylori activity. Archives of medical research. 1 Jul 2010;**41**(5):324-331

[55] Muñiz AE. Dermatitis. WB Saunders, Philadelphia: In: Pediatric Emergency Medicine. Elsevier; 1 Jan 2008. pp. 859-870

[56] Mortz CG, Andersen KE, Dellgren C, Barington T, Bindslev-Jensen C. Atopic dermatitis from adolescence to adulthood in the TOACS cohort: Prevalence, persistence and comorbidities. Allergy: European Journal of Allergy & Clinical Immunology. 2015;**70**(7):836-845

[57] Thomsen SF. Atopic dermatitis: Natural history, diagnosis, and treatment. ISRN Allergy. 2014;**2014**:1-7

[58] Alangari AA, Morris K, Lwaleed BA, Lau L, Jones K, Cooper R, et al. Honey is potentially effective in the treatment of atopic dermatitis: Clinical and mechanistic studies. Immunity, Inflammation and Disease. 2017;**5**(2):190-199

[59] Al-Waili NS. Topical application of natural honey, beeswax and olive oil mixture for atopic dermatitis or psoriasis: Partially controlled, single-blinded study. Complementary Therapies in Medicine. 2003;**11**(4):226-234

[60] Al-Waili NS. Clinical and mycological benefits of topical application of honey, olive oil and beeswax in diaper dermatitis. Clinical Microbiology and Infection. 2005;**11**(2):160-163

[61] Rindt IK, Niculae M, Brudaşcă F. Antibacterial Activity of Honey and Propolis Mellifera against *Staphylococcus aureus*. Lucr Stiint: Univ Stiint Agric a Banat Timisoara Med Vet; 2007

[62] Mohammed HA, Al-Lenjawi B, Mendoza D. Seborrheic dermatitis treatment with natural honey. Wounds International. 2018;**9**(3):36-38

[63] American Diabetes Association (ADA). Diagnosis and classification of diabetes mellitus. Diabetes Care. 2004;**27**(Supplement 1):S5-S10

[64] Erejuwa OO, Sulaiman SA, Ab Wahab MS. Honey—A novel antidiabetic agent. International Journal of Biological Sciences. 2012;**8**(6):913

[65] Erejuwa OO, Sulaiman SA, Wahab MSA. Oligosaccharides might contribute to the antidiabetic effect of honey: A review of the literature. Molecules. 2011;**17**(1):248-266

[66] Bobiş O, Dezmirean DS, Moise AR. Honey and diabetes: The importance of natural simple sugars in diet for preventing and treating different type of diabetes. Oxidative Medicine and Cellular Longevity. 2018:1-12

[67] Abdulrhman M, El Hefnawy M, Ali R, Abdel Hamid I, Abou El-Goud A, Refai D. Effects of honey, sucrose and glucose on blood glucose and C-peptide in patients with type 1 diabetes mellitus. Complementary Therapies in Clinical Practice. 2013;**19**(1):15-19

[68] Fasanmade AA, Alabi OT. Differential effect of honey on selected variables in alloxan-induced and fructose-induced diabetic rats. African Journal of Biomedical Research. 2008;**11**:191-196

[69] Al-Waili NS. Natural honey lowers plasma glucose, C-reactive protein, homocysteine, and blood lipids in healthy, diabetic, and hyperlipidemic subjects: Comparison with dextrose and sucrose. Journal of Medicinal Food. 2004;**7**(1):100-107

[70] Tsiapara AV, Jaakkola M, Chinou I, Graikou K, Tolonen T, Virtanen V, et al. Bioactivity of Greek honey extracts on breast cancer (MCF-7), prostate cancer (PC-3) and endometrial cancer (Ishikawa) cells: Profile analysis of extracts. Food Chemistry. 2009;**16**(3):702-708

[71] Fauzi AN, Norazmi MN, Yaacob NS. Tualang honey induces apoptosis and disrupts the mitochondrial membrane potential of human breast and cervical cancer cell lines. Food and Chemical Toxicology. 2011;**49**(4):871-878

[72] Jaganathan SK, Mandal M. Involvement of non-protein thiols, mitochondrial dysfunction, reactive oxygen species and p53 in honey-induced apoptosis. Investigational New Drugs. 2010;**28**(5):624-633

[73] Fernandez-Cabezudo MJ, El-Kharrag R, Torab F, Bashir G, George JA, El-Taji H, et al. Intravenous administration of manuka honey inhibits tumor growth and improves host survival when used in combination with chemotherapy in a melanoma mouse model. PLoS One. 2013;**8**(2):e55993

[74] Tomasin R, Cintra Gomes-Marcondes MC. Oral administration of Aloe vera and honey reduces walker tumour growth by decreasing cell proliferation and increasing apoptosis in tumour tissue. Phytotherapy Research. 2011;**25**(4):619-623

[75] Jaganathan SK. Honey constituents and their apoptotic effect in colon cancer cells. Journal of ApiProduct and ApiMedical Science. 2009;**1**(2):29-36

[76] Pichichero E, Cicconi R, Mattei M, Muzi MG, Canini A. Acacia honey and chrysin reduce proliferation of melanoma cells through alterations in cell cycle progression. International Journal of Oncology. 2010;**37**(4):973-981

[77] Lee Y-J, Kuo H-C, Chu C-Y, Wang C-J, Lin W-C, Tseng T-H. Involvement of tumor suppressor protein p53 and p38 MAPK in caffeic acid phenethyl ester-induced apoptosis of C6 glioma cells. Biochemical Pharmacology. 2003;**65**(12):2281-2289

[78] Šimúth J, Bíliková K, Kováčová E, Kuzmová Z, Schroder W. Immunochemical approach to detection of adulteration in honey: Physiologically active royal jelly protein stimulating TNF-α release is a regular component of honey. Journal of Agricultural and Food Chemistry. 2004;**52**(8):2154-2158

[79] Tonks AJ, Cooper RA, Jones KP, Blair S, Parton J, Tonks A. Honey stimulates inflammatory cytokine production from monocytes. Cytokine. 2003;**21**(5):242-247

[80] Tonks A, Cooper RA, Price AJ, Molan PC, Jones KP. Stimulation of tnf-α release in monocytes by honey. Cytokine. 2001;**14**(4):240-242

[81] Tsutsui T, Hayashi N, Maizumi H, Huff J, Barrett JC. Benzene-, catechol-,

hydroquinone- and phenol-induced cell transformation, gene mutations, chromosome aberrations, aneuploidy, sister chromatid exchanges and unscheduled DNA synthesis in Syrian hamster embryo cells. Mutation Research/Fundamental and Molecular Mechanisms of Mutagenesis. 1997;**373**(1):113-123

[82] Saxena S, Gautam S, Maru G, Kawle D, Sharma A. Suppression of error prone pathway is responsible for antimutagenic activity of honey. Food and Chemical Toxicology. 2012;**50**(3-4):625-633

[83] Wang X-H, Andrae L, Engeseth NJ. Antimutagenic effect of various honeys and sugars against Trp-p-1. Journal of Agricultural and Food Chemistry. 2002;**50**(23):6923-6928

[84] Oršolić N, Knežević AH, Šver L, Terzić S, Bašić I. Immunomodulatory and antimetastatic action of propolis and related polyphenolic compounds. Journal of Ethnopharmacology. 2004;**94**(2-3):307-315

[85] Porcza L, Simms C, Chopra M. Honey and cancer: Current status and future directions. Diseases. 2016;**4**(4):30

[86] Borda LJ, Macquhae FE, Kirsner RS. Wound dressings: A comprehensive review. Current Dermatology Reports. 2016;**5**(4):287-297

[87] Jull AB, Rodgers A, Walker N. Honey as a topical treatment for wounds. Cochrane Database of Systematic Reviews. 2008;**4**:CD005083

[88] Sare JL. Leg ulcer management with topical medical honey. British Journal of Community Nursing. 2008;**13**(9):S22, S24, S26 Passim

[89] Dunford CE, Hanano R. Acceptability to patients of a honey dressing for non-healing venous leg ulcers. Journal of Wound Care. 2004;**13**(5):193-197

[90] Simon A, Traynor K, Santos K, Blaser G, Bode U, Molan P. Medical honey for wound care—still the "latest resort"? Evidence-based Complementary and Alternative Medicine. 2009;**6**(2):165-173

[91] Zbuchea A. Up-to-date use of honey for burns treatment. Annals of Burns and Fire Disasters. 2014;**27**(1):22-30

[92] Malik KI, Malik MN, Aslam A. Honey compared with silver sulphadiazine in the treatment of superficial partial-thickness burns. International Wound Journal. 2010;**7**(5):413-417

[93] Erejuwa O, Sulaiman S, Wahab M. Effects of honey and its mechanisms of action on the development and progression of cancer. Molecules. 2014;**19**(2):2497-2522

[94] Dold H, Witzenhausen R. Ein Verfahren zur Beurteilung der örtlichen inhibitorischen (keimvermehrungshemmenden) Wirkung von Honigsorten verschiedener Herkunft. Zeitschrift für Hygiene und Infektionskrankheiten. 1955;**141**(4):333-337

[95] Tosi EA, Ré E, Lucero H, Bulacio L. Effect of honey high-temperature short-time heating on parameters related to quality, crystallisation phenomena and fungal inhibition. LWT-Food Science and Technology. 2004;**37**(6):669-678

[96] Chen C, Campbell LT, Blair SE, Carter DA. The effect of standard heat and filtration processing procedures on antimicrobial activity and hydrogen peroxide levels in honey. Frontiers in Microbiology. 2012;**3**:1-8

[97] Al-Waili NS. Investigating the antimicrobial activity of natural

honey and its effects on the pathogenic bacterial infections of surgical wounds and conjunctiva. Journal of Medicinal Food. 2004;7(2):210-222

[98] Subramanian R, Umesh Hebbar H, Rastogi NK. Processing of honey: A review. International Journal of Food Properties. 2007;**10**(1):127-143

[99] Mandal MD, Mandal S. Honey: Its medicinal property and antibacterial activity. Asian Pacific Journal of Tropical Biomedicine. 2011;**1**(2):154-160

[100] Kwakman PHS, Zaat SAJ. Antibacterial components of honey. IUBMB Life. 2012;**64**(1):48-55

[101] Wang XH, Gheldof N, Engeseth NJ. Effect of processing and storage on antioxidant capacity of honey. Journal of Food Science. 2006;**69**(2):fct96-fct101

[102] Molan PC. The antibacterial activity of honey. Bee World. 1992;**73**(1):5-28

Chapter 4

Health Benefits of Honey

Abstract

In addition to being used as food, honey has been used as an alternative medicine for thousands of years. Honey has a great potential to be used as a medicine because it is not suitable for micro-organisms, it is very acidic and has a very high sugar content, which causes an osmotic effect that prevents the growth of some micro-organisms, moreover, in some honey, hydrogen peroxide is found, which has a strong antibacterial effect. However, properties and appearances of honey vary greatly according to the floral source in which the bee collects the nectar, so some honey also have a strong antioxidant and anti-inflammatory activity. Recently, there are several studies, mainly *in vitro*, that prove the effectiveness of honey for various medical purposes due to its components and its antibacterial, anti-inflammatory, antioxidant, antiviral, antifungal, and anticancer properties.

Keywords: anti-inflammatory, antioxidant, bee, cancer, medicine

1. Introduction

Honey is a compound widely used as a medicine and food source for thousands of years [1]. Several natural products that have been used as medicine have been replaced by modern pharmaceuticals, but recently they have returned to the world stage due to the growing public interest [2]. In ancient Egypt, beekeeping has been practiced for more than 4000 years, and honey has been used as a medicine in the treatment of wounds, ulcers, burns, abscesses, gastrointestinal diseases, inflammations, rigid joints, and even as a contraceptive method [1, 3]. In Asia, honey is recognized for its medicinal value since 2000 BC [1]. There are also references to different uses of honey in the Bible and in the Qur'an [1]. The ancient Greek Hippocrates, known as the father of modern medicine, used honey to clean wounds, gastrointestinal diseases, and ulcers [1, 3]. In Ancient Rome, honey was also prescribed alone or in combinations, often used to treat throat problems, pneumonia, and even snake bites [1].

The main components of honey are sugars, among which are predominantly fructose and glucose [4, 5]. However, there are other compounds in smaller quantities and very variable depending on the type of each honey, from the floral source where the bee collects the nectar, such as water and free amino acids [4, 5]. Among them, the most found is proline [4, 6]. Some specific enzymes are also found, the main enzymes of honey are invertase, amylase, and glucose oxidase, but other enzymes such as catalase and phosphatase [6–8]. Honey is also composed of organic acids that contribute to its characteristic flavor and are responsible for the excellent stability of honey against micro-organisms, for example, formic, acetic, butyric, oxalic, lactic, succinic, folic, malic, citric, and glycolic [6, 7]. Gluconic acid is considered one of the most important organic acids in honey; it is the product of catalytic oxidation of glucose oxidase, in this oxidation, hydrogen peroxide is also formed, which has a strong antibacterial effect [4–7].

Honey may still have some mineral substances, such as potassium, magnesium, sodium, calcium, phosphorus, iron, manganese, cobalt, and copper; studies show that honey can contain several types of minerals, but potassium is the most abundant in various types of honey [6, 8–10]. Carotenoids, flavones, and anthocyanins can still be found, which contribute to the antioxidant action of honey [6]. About 80 aromatic compounds have been detected in honey, including carboxylic acids, aldehydes, ketones, alcohols, hydrocarbons, and phenols [6]. These compounds also contribute to the organoleptic properties of honey. The appearance of honey varies from almost colorless to dark brown; it can be liquid, viscous, or solid. Its flavor, aroma, and composition vary enormously, depending on the floral source in which the honeybee collects the nectar. However, some environmental factors can strongly influence honey composition, such as temperature and humidity [6, 7, 11].

Honey is a food that contains high energy carbohydrates, being that 95–99% of the total solids are composed by sugars, which are easily digestible, since they are similar to many fruits [7, 12]. Proteins and enzymes in honey often have no significant nutritional value, as they are usually not present in sufficient amounts [7]. Several of the essential vitamins are present in honey, such as vitamin K, B1, B2, B6, and C, but generally at insignificant levels [7, 8, 13]. The mineral content of honey is variable, usually darker honeys have significant amounts of minerals, but honey can be considered a nutritive sweetener, mainly due to its high fructose content [7, 13].

In addition to its food value, honey has great potential in medicine; it has been used for thousands of years, and has now been widely studied as an alternative medicine. Honey is not a suitable medium for bacteria, since it is very acidic and has a very high sugar content. This causes an osmotic effect that prevents the growth of bacteria, this effect works literally drying the bacteria [7, 13]. Another type of antibacterial property of honey was called inhibition in 1940 by Dold [7]. And in 1963, Jonathan White proposed that this inhibitory effect described in 1940 was due to the hydrogen peroxide produced and accumulated in the diluted honey, which we know today, is a by-product of the formation of gluconic acid by the enzyme glucose oxidase [5, 7, 11].

Historically, honey has been used for various medical purposes; and recent research has confirmed the effectiveness in the treatment of several diseases due to its components and its properties antibacterial, anti-inflammatory, antioxidants, antiviral, and others that will be addressed in this chapter.

2. Properties of honey

2.1 Anti-inflammatory

Inflammation is nothing more than a defense response of the body to a tissue that has suffered a certain damage, which consists of the recruitment of leucocytes and plasma proteins of the blood [14, 15]. This damage can be caused by physical, chemical, or even microbial agents; inflammation is characterized by edema, erythema, pain, and increased temperature [15, 16].

It is well known that propolis, another product from honeybee colony, has potential anti-inflammatory properties, including *in vivo.* But studies on the anti-inflammatory power of honey also are promising, such as the study that evaluated the anti-inflammatory and antioxidant effects of Tualang honey against conventional treatment in alkaline lesions in the eyes of rabbits and the results showed that there was no difference in the clinical inflammatory characteristics between the group treated with honey and the group with conventional treatment, so it is possible to infer that Tualang may be an alternative treatment [17, 18]. Other studies have also been depending on the use of honey, such as chronic ocular surface diseases and infectious conjunctivitis [19, 20].

Gastric ulcers are among the most common diseases affecting humans, a study demonstrated that the use of honey in conjunction with other compounds may promote gastroprotection. Later, a recent study investigated the effect of gastric protection using only honey against gastric ulcers induced by ethanol in rats and also suggested this effect as gastroprotection [21, 22]. Manuka honey significantly decreased the ulcer, completely protected the mucus of the lesions and preserved the gastric mucus glycoprotein, significantly increased the mucus levels of gastric nitric oxide, reduced glutathione, glutathione peroxidase, and superoxide dismutase, and also decreased lipid peroxidation of the mucus and tumor necrosis factor-α, interleukins-1β, and concentrations of interleukins-6 [21]. Honey has been shown to be efficient in other types of ulcers, and this Manuka honey exerted an antiulcer effect, keeping enzymes and antioxidants, non-enzymatic and inflammatory cytokines reduced [21, 23].

In addition to the Manuka honey and the Tualang honey, the anti-inflammatory effect of Malaysia's Gelam honey was also studied, which is associated with anti-inflammatory effects on tissues [24, 25]. Malaysia Gelam honey was tested in rats induced by inflammation [25]. Paw edema was induced by a subplantar injection and the rats were treated with either the anti-inflammatory drug Indomethacin or Gelam honey. Results showed that Gelam honey can reduce dose-dependent edema in inflamed rat paws, decrease the production of nitric oxide, prostaglandin, tumor necrosis factor-α, and interleukin-6 in plasma, and suppress expression of synthase inducible nitric oxide, cyclooxygenase-2, tumor necrosis factor-α, and interleucine-6 in paw tissue [25]. The oral pre-treatment of Gelam honey at 2 g/kg body weight at two times (1 and 7 days) showed a decreased production of proinflammatory cytokines, which was similar to the effect of the anti-inflammatory indomethacin, both in plasma and in the tissue, and Gelam honey has anti-inflammatory effects and is potentially useful for the treatment of inflammatory conditions [25]. Another study demonstrated that different types of honey promoted increased release of TNF-α, IL-1β, and IL-6 from monocytes, which are cells that assist in healing [26].

We can also compare the anti-inflammatory activity of honey with another herbal remedy in a study carried out in 2012 to test the activity of honey and brown sugar, surgically treated guinea pigs that were treated with honey, brown sugar, and a control group treated with saline solution, it is already known that sugar can help healing [27, 28]. The honey group showed a decrease in the area of the wound and the formation of granulation tissue before the brown sugar group and control; the honey group was still the only one that presented no crust in any wound and promoted a faster healing by stimulating the faster formation of granulation tissue and re-epithelization [28]. In addition, honey showed a higher antibacterial effect in relation to brown sugar and control group [28]. Another study had the same result, honey was effective in reducing bacterial contamination and wound healing [29].

Recent studies proved the anti-inflammatory activity of honey; different types of honey, different regions and different floral sources, were studied and both showed anti-inflammatory responses [17, 21, 25, 28]. Treatment with Tualang honey and Gelam honey showed similar responses to conventional anti-inflammatories used for specific treatments [17, 25]. Honey still has a better anti-inflammatory activity than brown sugar, promoting faster healing [28]. Also, honey is a relatively cheap and easily accessible anti-inflammatory compound that needs to be further studied and later applied in modern medicine [17, 21, 25, 28].

2.2 Antibacterial

One of the advances of modern medicine has been the development of antibiotics; these antibiotics can be bactericidal, which kill the micro-organisms directly, or bacteriostatic, which prevent the growth of micro-organisms [30]. However, micro-organisms are increasingly developing resistance to these antibiotics, which is a major concern. In addition to antibiotics, the prevention of bacterial diseases can be carried out with the use of vaccines and with basic sanitary methods [30, 31].

Many different micro-organisms can cause disease and be transmitted even by contaminated water, and among the major aquatic pathogens are *Escherichia coli* and *Pseudomonas aeruginosa.* Some studies have already shown that honey can combat these pathogens [14, 18, 32, 33]. A study in 2011 tested the bacterial activity of honey, for which the Revamil® and Manuka honey were used, and it was found that both honeys had activity against *Escherichia coli*, *Pseudomonas aeruginosa*, and also against *Bacillus subtilis* [34]. Manuka honey still had a greater efficacy than Revamil® against *Staphylococcus aureus-methicillin resistant* bacteria after 24-h incubation [34]. Despite the efficiency of honey, propolis has higher antibacterial activity against *Staphylococcus aureus* [35]. Overall, Revamil® honey clearly had more potent bactericidal activity than Manuka after 2 h of incubation, while Manuka honey was more potent after 24 h [34].

The bacteria *Streptococcus pyogenes* and *Streptococcus pneumoniae* are important human respiratory pathogens; *Streptococcus pneumoniae* can cause invasive lung infections that can develop in secondary infections and other respiratory disorders [14]. The antibacterial activity of honey was tested using dressings soaked with two types of honey, including Aquacel-Tualang honey and Aquacel-Manuka honey, the conventional dressing for burn treatment, Aquacel-Ag and only the curative Aquacel (control), against bacteria isolated from patients with burns (*in vitro*) [30]. Seven organisms were isolated from burns, four types of Gram-negative bacteria, *Enterobacter cloacae*, *Klebsiella pneumoniae*, *Pseudomonas* spp., and *Acinetobacter* spp., and three Gram-positive bacteria, *Staphylococcus aureus*, Coagulase-negative *Staphylococcus aureus*, and *Streptococcus* spp. Aquacel-Ag and Aquacel-Manuka dressings provided a better zone of inhibition for Gram-positive bacteria. However, similar results between Aquacel-Manuka and Aquacel-Tualang were obtained against Gram-negative bacteria [36].

Salmonellosis is a gastrointestinal disease caused by eating food contaminated with *Salmonella*, such as eggs, chicken, meat, and raw vegetables, or by handling animal or animal products contaminated by the bacterium [14, 37]. It is the most common bacterial food infection in the United States. However, most *Escherichia coli* strains are not pathogenic to humans, but the few pathogenic strains of *Escherichia coli* are transmitted by food and produce potent enterotoxins [14]. In the literature, there are several studies that demonstrate the efficiency of honey against bacteria important to human health, one of them demonstrated the antibacterial potential of honey against clinical isolates of *Escherichia coli*, *Pseudomonas aeruginosa*, and *Salmonella enterica Typhi* by *in vitro* methods [38]. Honey showed excellent antibacterial activity against all bacteria studied, which are related, respectively, to urinary tract infection, skin lesion, and enteric fever in human patients; and thus, honey can be considered an alternative treatment against such infection [38]. In addition to honey being effective against bacterial infections, it can be used as a treatment for one of the most common bacterial contamination symptoms, when honey is administered as oral rehydration fluid, it can decrease the duration of bacterial diarrhea [39]

Another form of food poisoning is caused by enterotoxins produced by Gram-positive bacteria, such as *Staphylococcus aureus*; these toxins cause nausea, vomiting, diarrhea, and dehydration, and is a major public health problem [14, 40]. The antibacterial action of Tualang, Gelam, and Durian honeys was tested against *Staphylococcus aureus*, *Staphylococcus epidermidis*, *Enterococcus faecium*, *Enterococcus faecalis*, *Escherichia coli*, *Salmonella enterica Typhi*, and *Klebsiella pneumoniae* [41]. Durian honey did not produce substantial antibacterial activity, while Tualang and Gelam honey showed a spectrum of antibacterial activity with its growth inhibitory effects against all bacterial species tested, including vancomycin-resistant *Enterococci* (VRE), the results still suggest the Gelam honey has the highest antibacterial effect among the honey samples from Malaysia tested [41].

Clostridiums are anaerobic bacteria that are capable of growing up in canned food [14]. In addition to the antibacterial activity of honey against the bacteria dating to the top, Manuka honey still has antibacterial effect on *Clostridium difficile*, which is a Gram-positive anaerobic bacillus, which was associated with approximately 29,000 deaths in 2011 in the United States [42, 43]. A recent study has shown that Manuka honey exhibited a bactericidal action against *Clostridium difficile*; this is yet another feature that makes Manuka honey highly attractive in the treatment of bacterial infections [42]. However, Manuka honey was considered ineffective against other bacteria *Helicobacter pylori* when tested *in vivo*, despite having been found effective *in vitro* [44, 45].

Honey has an excellent antibacterial effect against different types of bacteria, as previously mentioned; honey is very acidic and has a very high sugar content, which does not serve as a suitable medium for bacteria [4–7]. Moreover, in some honeys, the peroxide of hydrogen is found, which has a strong antibacterial effect [4–7]. Remavil® honeys, Manuka honey, Tualang honey, and Gelam honey were tested with different types of bacteria and had positive results [34, 36, 41, 42]. The bacteria tested and susceptible to some of these honeys were *Escherichia coli*, *Pseudomonas aeruginosa*, *Pseudomonas* spp., *Bacillus subtilis*, *Staphylococcus aureus*, *Staphylococcus aureus-resistant methicillin*, *coagulase-negative Staphylococcus aureus*, *Staphylococcus epidermidis*, *Enterobacter cloacae*, *Klebsiella pneumoniae*, *Acinetobacter* spp., *Streptococcus* spp., *Enterococcus faecium*, *Enterococcus faecalis*, *Salmonella enterica serovar Typhimurium*, *vancomycin-resistant Enterococci*, and *Clostridium difficile* [34, 36, 38, 41, 42].

2.3 Antivirals

Of all human infectious diseases, the most prevalent and difficult to treat are those that are caused by viruses, because viruses usually remain infectious in dry mucus for a long time [14]. Also, viruses need a host cells to occur its replication; so killing the virus means killing your host cell as well. Hence, vaccination is the most efficient way to prevent these diseases [14, 46].

Chickenpox is caused by the varicella-zoster virus and it is a very common childhood disease that usually does not cause many problems; but when it affects the elderly, it can be easily fatal [14, 47]. Varicella-zoster is highly contagious and is transmitted by infectious droplets, which results in a systemic rash on the skin [14]. As honey can be conveniently applied to the skin, it is easily found and relatively inexpensive, it can be considered an excellent remedy against Zoster rash, especially in developing countries, or in countries where antiviral drugs are relatively expensive and difficult to access. Therefore, a study determined *in vitro* antiviral effect of honey against the varicella-zoster virus; two types of honey were used, Manuka honey and clover honey, and both types showed antiviral activity against the varicella-zoster virus, showing that honey has significant antiviral activity against varicella-zoster [48]. A study on the relationship of honey to another virus, analyzed *in vivo*, showed that the use of topical honey is safe and effective in the treatment of recurrent herpes and genital herpes lesions [49].

Respiratory syncytial virus is the most common cause of viral respiratory infections in infants and young children, also seriously affects adults, the elderly and immunocompromised, causing deaths mainly in the elderly [50, 51]. The antiviral activity of honey was tested for its action against the respiratory syncytial virus. A variety of tests using cell culture was developed to assess the susceptibility of respiratory syncytial virus to honey. The results confirmed that treatment with honey promoted inhibition of viral replication [50]. Attempts to isolate the antiviral component in honey demonstrated that sugar was not responsible for the inhibition of respiratory syncytial virus, but could be methylglyoxal; this component of honey may play a role in the increased potency of Manuka honey against respiratory syncytial virus [50]. Thus, honey may be an alternative and effective antiviral treatment for the therapy of respiratory viral infections, such as respiratory syncytial virus; however, other measures, such as an effective vaccine, are still necessary for the control of this disease [50, 52].

Influenza is a highly infectious respiratory disease of viral origin that causes even more deaths than the respiratory syncytial virus at all ages, except in children less than a year old [14, 51]. Influenza viruses are transmitted from person to person through the air, especially from droplets expelled during coughing and sneezing and are a serious threat to human health, and there is an urgent need for the development of new drugs against these viruses. Therefore, the anti-influenza virus activity of honey from several sources was studied [53]. The results showed that honey, in general, and particularly Manuka honey, has potent inhibitory activity against the influenza virus, demonstrating a potential medicinal value [53]. In addition to honey, propolis has also been studied against the influenza virus and appears to decrease the activity of the influenza virus [54].

Honey, especially Manuka honey, has strong antiviral properties. Studies show that honey has action against the varicella-zoster virus, the respiratory syncytial virus, and also has anti-influenza activity [47, 50, 53]. New studies on this property of honey are necessary, mainly with other types of honey.

2.4 Antifungal

Most people associate fungi with organic matter decomposition or superficial fungal infections, but fungi can cause various human diseases, from mild to firmly established systemic diseases; the most serious infections can even be fatal [14]. The incidence of *Candida* infections is increasing worldwide. *Candida albicans* is present in the normal human microbiota; however, this fungus can cause a variety of diseases, such as vaginal, oral, and systemic infections, especially in immunosuppressed patients, as carriers of the HIV virus, these infections can be further aggravated by the increase in resistance levels of this fungus to the medicines [14, 55, 56]. Clinical isolates of *Candida albicans*, *Candida glabrata*, and *Candida dubliniensis* were tested against four different honeys. The antifungal activities of floral honeys were significantly higher than artificial honey against *Candida albicans* and *Candida glabrata*; but for *Candida dubliniensis*, only Jarrah honey was significantly active [56]. *Candida glabrata*, which is innate less susceptible to many conventional antifungals, was also the least susceptible to the honey tested [56].

As previously stated, honey has antifungal properties and may act against *Candida* [57]. A study in 2012 evaluated the clinical and mycological cure rates of a mixture of honey and vaginal mucus compared to local antifungal agents for the treatment of patients with vulvovaginal candidiasis during pregnancy, recurrent asymptomatic candidiasis in early pregnancy is associated with preterm birth [57, 58]. The clinical cure rate was significantly higher in the honey and mucus group than in the conventional antifungal group, while the mycological cure rate was higher in the conventional antifungal group than in the mucus and honey group; therefore, the mixture of honey and mucus can be used with a complement or an alternative to antifungal agents, especially in patients with vulvovaginal candidiasis during pregnancy [57].

In addition to the antifungal activity of honey against *Candida albicans*, the antifungal activity against *Rhodotorula* sp. was studied; this fungus can also affect humans, cases of meningitis caused by *Rhodotorula* species in immunosuppressed people have been reported [59, 60]. Four honeys from Algeria from different botanical origins were analyzed to test the antifungal effect against *Candida albicans* and *Rhodotorula* sp., different concentrations of honey were studied *in vitro* for antifungal activity, and the study demonstrated that, *in vitro*, these natural products clearly show antifungal activity against *Rhodotorula* sp. and *Candida albicans* [60].

Aspergillus spp. is a saprophyte commonly found in nature as a mold of leaves, produces potent allergens, and often causes asthma and other hypersensitivity reactions [14]. The antifungal activities of some samples of honey obtained from different geographic locations in Nigeria were tested against some fungal isolates [61]. Honey samples were examined for antifungal activity against *Aspergillus niger*, *Aspergillus flavus*, *Penicillium chrysogenum*, *Microsporum gypseum*, *Candida albicans*, and *Saccharomyces* sp., and results show that honey samples had different levels of inhibitory activity at various concentrations against the fungi tested, with zones of inhibition increasing with increasing honey concentration; *Microsporum gypseum*, which can infect immunosuppressed patients, was the most sensitive of all fungal isolates studied, while *Candida albicans* was the least sensitive, other studies have shown efficient inhibitory activity of honey against the growth of *Candida albicans* [61–64]. Honey samples used in the study showed spectrum and promising antifungal activity, the honey from Nigeria may serve as a source of antifungal for possible development of antifungal drugs for the treatment of fungal infections [61].

Besides the antibacterial and antiviral properties, some honeys also have antifungal properties [56, 57, 59, 61]. Recent studies showed some honey have properties against *Candida albicans*, *Candida glabrata*, *Candida dubliniensis*, *Rhodotorula* sp., *Aspergillus niger*, *Aspergillus flavus*, *Penicillium chrysogenum*, *Microsporum gypseum*, and *Saccharomyces* sp., which make these honey as possible alternative medicines, especially against candidiasis, a disease that is growing worldwide [24, 56, 59, 61].

2.5 Anticancer

In 2016, the cancer mortality rate has dropped 23% since 1991 [65]. Despite this progress, mortality rates are increasing for liver, pancreatic, and uterine cancers; and cancer is now the leading cause of death in 21 states from United States, lung cancer is still the most lethal, followed by breast cancer [65, 66]. The advance for cancer treatment needs more clinical and basic research [65].

Many scientists have focused on the antioxidant property of honey. Studies indicate that ingestion of honeybee products, such as honey, can prevent cancer [67, 68]. Through the use of human renal cancer cells, the antiproliferative activities, apoptosis, and the antitumor activity of honey were investigated [67]. Honey decreased cell viability in malignant cells regardless of concentration and time [67]. Honey induced apoptosis of human renal cancer cells according to honey concentration, and apoptosis plays an important role, most of the drugs used in the treatment of cancer are apoptotic inducers, so the apoptotic nature of honey is considered vital [67].

The anticancer activity of honey samples was extracted from three different Egyptian floral sources and was tested against colon, breast, and liver tumor lineage [69]. Cassia honey showed moderate cytotoxic activity against colon cancer and breast cancer, with the weakest cytotoxic activity against liver cancer; Citrus honey exhibited the highest cytotoxic activity against breast cancer; and Ziziphus honey showed potent efficiency against colon, liver, and breast cancer [69]. Breast cancer, which is the type of cancer that most affects and kills women, was also tested for another type of honey, the Manuka honey, and the results showed that it is cytotoxic to MCF-7 breast cancer cells *in vitro* and the effects are mainly correlated with the total content of phenols and their antioxidant power [65, 70].

The phytochemical content and antioxidant activity of melon honey and Manuka honey and their cytotoxic properties were tested against human and metastatic colon adenocarcinoma. The ability to induce apoptosis in colon cancer cells depends on the concentration of honey and type of cell line, in addition to having a great relation with the phenolic content and residues of tryptophan. Honey was analyzed for phenolic, flavonoid, amino acid, and protein contents, as well as their free radical scavenging activities [71, 72]. Melon honey presented the highest amount of phenolics, flavonoids, amino acids, and proteins, as well as antioxidant capacity in relation to Manuka honey [71]. Both melon honey and Manuka honey induced cytotoxicity and cell death independently of dose and time in human and metastatic colon adenocarcinoma cells [71]. Melon honey showed to be more efficient in concentrations [71]. The results indicate that melon honey and Manuka honey can induce inhibition of cell growth and the generation of reactive oxygen species in colon adenocarcinoma and metastatic cells, which may be due to the presence of phytochemicals with antioxidant properties. These results suggest a potential chemo-preventive agent against colon cancer; in addition, honey can improve the functioning of other substances already used in cancer treatment [71, 73].

Research on cancer control has shown the importance of adjuvant therapies [74]. *Aloe vera* may reduce tumor mass and rates of metastasis, and its association with conventional therapy can produce benefits for the treatment, while honey may inhibit tumor growth [74, 75]. The influence of *Aloe vera* and honey on tumor growth and the apoptosis process was evaluated by evaluating tumor size, the rate of cell proliferation for Walker 256 carcinoma [74]. Tumor-bearing mice received a daily dose of *Aloe vera* and honey, and the control group received only sodium chloride solution [74]. The effect of *Aloe vera* and honey against tumor growth was observed through a decrease in relative weight (%) [74]. The results suggested that *Aloe vera* and honey can modulate tumor growth, reduce cell proliferation, and increase susceptibility to apoptosis. Studies have shown that honey has antiproliferative activity because of its ability to induce apoptosis, so this combination is a possible adjuvant therapy [74, 76, 77].

Several types of honey have been studied because of their anticancer properties [65, 67, 69–71, 74]. Currently, cancer is one of the world's leading diseases, requiring further studies [65]. Some honey have already been tested against colon, breast, and liver tumor, as well as human kidney cancer and Ehrlich ascites carcinoma cell lines, where most have weak to strong cytotoxic activity depending on the type of honey tested and depending on the dose of honey [67, 69–71] The effect of *Aloe vera* on honey has also been studied, and the whole has the capacity to modulate tumor growth, reducing cell proliferation, and also increasing susceptibility to apoptosis [74]. The antitumor effects of honey were highly correlated with their ability to induce apoptosis of cells and with their antioxidant power [65, 67, 69–71, 74]. The effect of

Aloe vera along with honey has also been studied, and the set has the capacity to modulate tumor growth, reducing cell proliferation, and also increasing susceptibility to apoptosis [74]. The antitumor effects of honey were highly correlated with its ability to induce cell apoptosis and with its antioxidant activity [65, 67, 69–71, 74].

2.6 Antioxidants

Antioxidants, which are present in large amounts of honey, making it a food with great antioxidative potential, are free radical scavengers that reduce the formation or neutralize free radicals [11, 78]. A comparative analysis of total phenolic content and antioxidant potential of commercially available common honey was performed along with Malaysia's Tualang honey. Biochemical analyzes revealed a significantly high phenolic content in Tualang honey [78]. In addition, the antioxidant capacity of Tualang honey was higher than that of common honey; these data suggested that the high activity of elimination of free radicals and antioxidant activity observed in Tualang honey were due to the increase in the level of phenolic compounds, it was also observed that the antioxidant activity of honey depends on its botanical origin [78, 79]. Therefore, the favorable antioxidant properties of Tualang honey can be important for nutrition and human health [78].

Type 2 diabetes consists of progressive hyperglycemia, insulin resistance, and β-pancreatic cell failure, which may result from glucose toxicity, inflammatory cytokines, and oxidative stress, and is responsible for 90–95% of all cases of diabetes [80, 81]. A study investigated the effect of pre-treatment with Gelam honey, and the individual flavonoid components chrysin, luteolin, and quercetin on the production of reactive oxygen species, cell viability, lipid peroxidation, and insulin in hamster pancreatic cells, cultured under normal conditions and hyperglycemic, the pre-treatment of cells with Gelam honey extract or flavonoid components showed a significant decrease in the production of reactive oxygen species, glucose-induced lipid peroxidation, and a significant increase in insulin content and viability of cultured cells under hyperglycemic conditions. The results indicated the *in vitro* antioxidant property of Gelam honey and flavonoids on hamster β cells, creating a protective effect against hyperglycemia [80]. Another study demonstrated the effect of honey on diabetics, the study with rats concluded that the pancreatic tissues of rats with diabetes were exposed to great oxidative stress and that supplementation with other honey, Tualang honey, had protective effects in the pancreas [80, 82].

Honey contains antioxidants, such as phenolic compounds that prevent cellular oxidative damage that leads to aging, disease such as cancer, metabolic disturbances, cardiovascular dysfunction and even death [83, 84]. The antioxidant effect of honey in young and middle-aged rats was compared, the rats were fed with pure water (control), those supplemented with 2.5 and 5.0 g/kg of Gelam honey for 30 days. Results showed that Gelam honey supplementation reduced DNA damage, plasma malondialdehyde level, and glutathione peroxidase. Liver activity superoxide dismutase also decreased in young rats supplemented with 5 g/kg of Gelam honey [84]. Gelam honey reduces the oxidative damage of young and middle-aged rats by modulating the activities of the antioxidant enzymes that were more prominent in higher concentration compared to the lower concentration [84]. Another study indicates that honey has these antioxidant and free radical sequestering properties, mainly due to its phenolic compounds [85].

Honey has antioxidant properties that can be further explored and studied, because antioxidants reduce free radicals and oxidative stress, which can help to promote and maintain health [80, 82, 84]. Besides the previously described, the antioxidant effect of honey can be an important property to help in the anticancer effect [67, 71].

3. Conclusions

Several studies have proven the effectiveness of honey as an alternative medicine; some have even shown that honey is as good a medicine as conventional medicine. Use of different types of honeys showed anti-inflammatory effect very similar to the conventional drug and that can be used as an alternative medicine in the treatment of diseases or inflammations. Honey can also be used as an antimicrobial agent anti-inflammatory, antibacterial, antivirals, antifungal, anticancer, and antioxidants. However, there is still a need to increase research on honey, especially in its potential as a medicine and also a dissemination of this knowledge to the population and the medical community, so an increase in the use of this powerful compound will be possible.

Conflict of interest

The authors declare that there is no conflict of interest

References

[1] Jones R. Prologue: Honey and healing through the ages. Journal of Apiproduct and Apimedical Science. 2009;**1**:2-5. DOI: 10.3896/ibra.4.01.1.02

[2] Ghosh S, Playford RJ. Bioactive natural compounds for the treatment of gastrointestinal disorders. The Biochemical Society. 2003;**104**:547-556. DOI: 10.1042/CS20030067

[3] Sato T, Miyata G. The nutraceutical benefit, part iii: Honey. Nutrition. 2000;**16**:468-469. DOI: 10.1016/s0899-9007(00)00271-9

[4] Steeg E, Montag A. Minorbestandteile des Honigs mit Aroma-Relevanz. Deutsche Lebensmittel Rundschau. 1988;**84**:147

[5] White JW Jr, Subers MH, Schepartz AJ. The identification of inhibine, the antibacterial factor in honey, as hydrogen peroxide and its origin in a honey glucose-oxidase system. Biochimica et Biophysica Acta. 1963;**73**:57-70. DOI: 10.1016/0926-6569(63)90108-1

[6] Sikorski ZE. Chemical and Functional Properties of Food Saccharides. Boca Raton: CRC Press; 2004. 440 p

[7] White JW Jr, Doner LW. Honey composition and properties. Beekeeping in the United States Agricultural Handbook. 1980;**335**:82-91

[8] Bogdanov S, Jurendic T, Sieber R, Gallmann P. Honey for nutrition and health: A review. Journal of the American College of Nutrition. 2008;**27**:677-689. DOI: 10.1080/07315724.2008.10719745

[9] Vanhanen LP, Emmetz A, Savage GP. Mineral analysis of mono-floral New Zealand honey. Food Chemistry. 2011;**128**:236-240. DOI: 10.1016/j.foodchem.2011.02.064

[10] Chua LS, Abdul-Rahaman NL, Sarmidi MR, Aziz R. Multi-elemental composition and physical properties of honey samples from Malaysia. Food Chemistry. 2010;**135**:880-887. DOI: 10.1016/j.foodchem.2012.05.106

[11] Gheldof N, Wang XH, Engeseth NJ. Identification and quantification of antioxidant components of honeys from various floral sources. Journal of Agricultural and Food Chemistry. 2002;**50**:5870-5877

[12] Doner LW. The sugars of honey—A review. Journal of the Science of Food and Agriculture. 1977;**28**:443-456. DOI: 10.1002/jsfa.2740280508

[13] White JW Jr. Honey. Advances in Food Research. 1978;**24**:287-374. DOI: 10.1016/s0065-2628(08)60160-3

[14] Madigan MT, Martinko JM, Bender KS, Buckley DH, Stahl DA. Microbiologia de Brock. 14th ed. Artmed: Porto Alegre; 2010. 1006 p

[15] Abbas AK, Lichtman AH, Pillai S. Imunologia Celular e Molecular. 9th ed. Rio de Janeiro: Elsevier; 2019. 516 p

[16] Roitt IM, Delves PJ, Burton DR, Martin SJ. Fundamentos de Imunologia. 13th ed. Rio de Janeiro: Guanabara Koogan; 2013. 544 p

[17] Bashkaran K, Zunaina E, Bakiah S, Sulaiman SA, Sirajudeen K, Naik V. Anti-inflammatory and antioxidant effects of Tualang honey in alkali injury on the eyes of rabbits: Experimental animal study. BMC Complementary and Alternative Medicine. 2011;**11**:1-11. DOI: 10.1186/1472-6882-11-90

[18] Khayyal MT, El-ghazaly MA, El-khatib AS. Mechanisms involved in the antiinflammatory effect of propolis extract. Drugs Under Experimental And Clinical Research. 1993;**19**:197-203

[19] Ilechie A, Kwapong PK, Mate-Kole E, Kyei S, Darko-Takyi. The efficacy of stingless bee honey for the treatment of bacteria-induced conjunctivitis in guinea pigs. Journal of Experimental Pharmacology. 2012;**4**:63-68. DOI: 10.2147/jep.s28415

[20] Albietz JM, Lenton LM. Effect of antibacterial honey on the ocular flora in tear deficiency and meibomian gland disease. Cornea. 2006;**25**:1012-1019. DOI: 10.1097/01.ico.0000225716.85382.7b

[21] Almasaudi SB, El-Shitany NA, Abbas AT, Abdel-Dayem UA, Ali SS, Al Jaouni SK, et al. Antioxidant, anti-inflammatory, and antiulcer potential of manuka honey against gastric ulcer in rats. Oxidative Medicine and Cellular Longevity. 2016;**2016**:1-10. DOI: 10.1155/2016/3643824

[22] Nasuti C, Gabbianelli R, Falcioni G, Cantalamessa F. Antioxidative and gastroprotective activities of anti-inflammatory formulations derived from chestnut honey in rats. Nutrition Research. 2006;**26**:130-137. DOI: 10.1016/j.nutres.2006.02.007

[23] Mohamed H, Salma MA, Al Lenjawi B, Abdi S, Gouda Z, Barakat N, et al. The efficacy and safety of natural honey on the healing of foot ulcers: A case series. Wounds. 2015;**27**:103-114

[24] Kassim M, Achoui M, Mansor M, Yussof KM. The inhibitory effects of Gelam honey and its extracts on nitric oxide and prostaglandin E2 in inflammatory tissues. Fitoterapia. 2010;**81**:1196-1201. DOI: 10.1016/j.fitote.2010.07.024

[25] Hussein SZ, Yussof KM, Makpol S, Yusof YAM. Gelam honey inhibits the production of proinflammatory, mediators NO, PGE2, TNF-α, and IL-6 in carrageenan-induced acute paw edema in rats. Evidence-based Complementary and Alternative Medicine. 2012;**2012**:1-13. DOI: 10.1155/2012/109636

[26] Tonks A. Honey stimulates inflammatory cytokine production from monocytes. Cytokine. 2003;**21**:242-247. DOI: 10.1016/s1043-4666(03)00092-9

[27] Cavazana WC, Simoes MLPB, Yoshii SO, Amado CAB, Cuman RKN. Açúcar (sacarose) e triglicerídeos de cadeia média com ácidos graxos essenciais no tratamento de feridas cutâneas: Estudo experimental em ratos. Anais Brasileiros de Dermatologia. 2009;**3**:229-236

[28] Santos IFC, Grosso SLS, Bambo OB, Nhambirre AP, Cardoso JMM, Schmidt SMS, et al. Mel e açúcar mascavo na cicatrização de feridas. Ciência Rural. 2012;**42**:2219-2224. DOI: 10.1590/s0103-84782012001200018

[29] Mphande AN, Killowe C, Phalira S, Jones HW, Harrison WJ. Effects of honey and sugar dressings on wound healing. Journal of Wound Care. 2007;**16**:317-319. DOI: 10.12968/jowc.2007.16.7.27053

[30] Tortora GJ, Funke BR, Case CL. Microbiologia. 10th ed. Artmed: Porto Alegre; 2012. 934 p

[31] Pommerville JC. Alcamo's Fundamentals of Microbiology. Jones and Bartlett Learning: Burlington; 2010. 860 p

[32] Kumar P, Sindhu RK, Narayan S, Singh I. Honey collected from different floras of Chandigarh tricity: A comparative study involving physicochemical parameters and biochemical activities. Journal of Dietary Supplements. 2010;**7**:303-313. DOI: 10.3109/19390211.2010.508034

[33] Agbaje EO, Ogunsanya T, Aiwerioba OIR. Conventional use of honey as antibacterial agent. Annals of African Medicine. 2006;**5**:78-81

[34] Kwakman PHS, Velde AA, Boer L, Vandenbroucke-Grauls CMJE, Zaat SAJ. Two major medicinal honeys have different mechanisms of bactericidal activity. PLoS One. 2011;**6**:17709-17709. DOI: 10.1371/journal.pone.0017709

[35] Miorini PL, Levy Junior NC, Custodio AR, Bretz WA, Marcucci MC. Antibacterial activity of honey and propolis from *Apis mellifera* and *Tetragonisca angustula* against *Staphylococcus aureus*. Journal of Applied Microbiology. 2003;**95**:913-920. DOI: 10.1046/j.1365-2672.2003.02050.x

[36] Nasir NA, Halim SA, Singh KK, Dorai AA, Haneef MN. Antibacterial properties of tualang honey and its effect in burn wound management: A comparative study. BMC Complementary and Alternative Medicine. 2010;**10**:1-7. DOI: 10.1186/1472-6882-10-31

[37] Pires SM, Vieira AR, Hald T, Cole D. Source attribution of human salmonellosis: An overview of methods and estimates. Foodborne Pathogens and Disease. 2014;**11**:667-676. DOI: 10.1089/fpd.2014.1744

[38] Mandal S, DebMandal M, Pal NK, Saha K. Antibacterial activity of honey against clinical isolates of *Escherichia coli*, *Pseudomonas aeruginosa* and *Salmonella enterica serovar Typhi*. Asian Pacific Journal of Tropical Medicine. 2010;**3**:961-964. DOI: 10.1016/s1995-7645(11)60009-6

[39] Haffejee IE, Moosa A. Honey in the treatment of infantile gastroenteritis. BMJ. 1985;**290**:1866-1867. DOI: 10.1136/bmj.290.6485.1866

[40] Klevens RM, Morrison MA, Nadle J, Petit S, Gershman K, Ray S, et al. Invasive methicillin-resistant *Staphylococcus aureus* infections in the United States. Journal of the American Medical Association. 2007;**298**:1763-1771. DOI: 10.1001/jama.298.15.1763

[41] Ng WJ, Ken KW, Kumar RV, Gunasagaran H, Chandramogan V, Lee YY. In-vitro screening of malaysian honey from different floral sources for antibacterial activity on human pathogenic bacteria. African Journal of Traditional, Complementary and Alternative Medicines. 2014;**11**:315-318. DOI: 10.4314/ajtcam.v11i2.14

[42] Hammond EN, Donkor ES. Antibacterial effect of Manuka honey on *Clostridium difficile*. BMC Research Notes. 2013;**6**:1-5. DOI: 10.1186/1756-0500-6-188

[43] Lessa FC, Mu Y, Bamberg WM, Beldavs ZG, Dumyati GK, Dunn JR, et al. Burden of *Clostridium difficile* infection in the United States. New England Journal of Medicine. 2015;**372**:825-834. DOI: 10.1056/nejmoa1408913

[44] McGovern DPB, Abbas SZ, Vivian G, Dalton HR. Manuka honey against *Helicobacter pylori*. Journal of the Royal Society of Medicine. 1999;**92**:439-439. DOI: 10.1177/014107689909200832

[45] Al Somal N, Coley KE, Molan PC, Hancock BM. Susceptibility of *Helicobacter pylori* to the antibacterial activity of manuka honey. Journal of the Royal Society of Medicine. 1994;**87**:9-12

[46] Morse SA, Butel JS, Brooks GF, Carrol KC, Mietzner-Amgh TA. Microbiologia médica de Jawetz, Melnick e Adelberg. 26th ed. AMGH: Porto Alegre; 2014

[47] Donahue JG, Choo PW, Manson JE, Platt R. The incidence of herpes zoster. Archives of Internal Medicine. 1995;**155**:7-21. DOI: 10.1001/archinte.1995.00430150071008

[48] Shahzad A, Cohrs RJ. In vitro antiviral activity of honey against varicella zoster virus (VZV): A translational medicine study for potential remedy for shingles.

International Archives of Medicine. 2012;**2**:1-7. DOI: 10.3823/434

[49] Al-Waili NS. Topical honey application vs. acyclovir for the treatment of recurrent herpes simplex lesions. Diagnostics and Medical Technology. 2004;**8**:94-98

[50] Zareie PP. Honey as an antiviral agent against respiratory syncytial virus [thesis]. Tauranga: University of Waikato; 2011

[51] Thompson WW, Shai DK, Weintrub E, Brammer L, Cox N, Anderson LJ, et al. Mortality associated with influenza and respiratory syncytial virus in the United States. Journal of the American Medical Association. 2003;**289**:179-190. DOI: 10.1001/jama.289.2.179

[52] Falsey AR, Hennessey PA, Formica MA, Cox C, Walsh EE. Respiratory syncytial virus infection in elderly and high-risk adults. New England Journal of Medicine. 2005;**352**:1749-1759. DOI: 10.1056/nejmoa043951

[53] Watanabe K, Rahmasari R, Matsunaga A, Haruyama T, Kobayashi N. Anti-influenza viral effects of honey in vitro: Potent high activity of manuka honey. Archives of Medical Research. 2014;**45**:359-365. DOI: 10.1016/j.arcmed.2014.05.006

[54] Shimizu T, Hino A, Tsusumi A, Park YK, Watanabe W, Kurokawa M. Anti-influenza virus activity of propolis in vitro and its efficacy against influenza infection in mice. Antiviral Chemistry and Chemotherapy. 2008;**19**:7-13. DOI: 10.1177/095632020801900102

[55] Sangeorzan JA, Bradley SF, He X, Zarins LT, Ridenour GL, Tiballi RN, et al. Epidemiology of oral candidiasis in HIV-infected patients: Colonization, infection, treatment, and emergence of fluconazole resistance. The American Journal of Medicine. 1994;**97**:339-346. DOI: 10.1016/0002-9343(94)90300-x

[56] Irish J, Carter DA, Shokohi T, Blair SE. Honey has an antifungal effect against Candida species. Medical Mycology. 2006;**44**:289-291. DOI: 10.1080/13693780500417037

[57] Abdelmonem AM, Rasheed SM, Mohamed AS. Bee-honey and yogurt: A novel mixture for treating patients with vulvovaginal candidiasis during pregnancy. Archives of Gynecology and Obstetrics. 2012;**286**:109-114. DOI: 10.1007/s00404-012-2242-5

[58] Farr A, Kiss H, Holzer I, Husslein P, Hangmann M, Petricevic L. Effect of asymptomatic vaginal colonization with *Candida albicanson* pregnancy outcome. Acta Obstetricia et Gynecologica Scandinavica. 2015;**94**:989-996. DOI: 10.1111/aogs.12697

[59] Moussa A, Noureddine D, Saad A, Abdelmelek M, Abdelkader B. Antifungal activity of four honeys of different types from Algeria against pathogenic yeast: *Candida albicans* and *Rhodotorula sp*. Asian Pacific Journal of Tropical Biomedicine. 2012;**2**:554-557. DOI: 10.1016/s2221-1691(12)60096-3

[60] Baradkar VP, Kumar S. Meningitis caused by *Rhodotorula mucilaginosain* human immunodeficiency virus seropositive patient. Annals of Indian Academy of Neurology. 2008;**11**:245-247. DOI: 10.4103/0972-2327.44561

[61] Anyanwu U. Investigation of in vitro antifungal activity of honey. Journal of Medicinal Plants Research. 2012;**6**:3512-3516. DOI: 10.5897/jmpr12.577

[62] Porro AM, Yoshioka MC, Kaminski SK, Palmeira MC, Fischman O, Alchorne MM. Disseminated dermatophytosis caused by *Microsporum gypseum* in two patients with the acquired immunodeficiency syndrome. Mycopathologia. 1997;**137**:9-12

[63] Theunissen F, Grobler S, Gedalia I. The antifungal action of three South African honeys on *Candida albicans* Apidologie. 2001;**32**:371-379. DOI: 10.1051/apido:2001137

[64] Al-waili NS. Mixture of honey, beeswax and olive oil inhibits growth of staphylococcus aureus and *Candida albicans*. Archives of Medical Research. 2005;**36**:10-13. DOI: 10.1016/j.arcmed.2004.10.002

[65] Siegel RL, Miller KD, Jemal A. Cancer statistics, 2016. CA: A Cancer Journal for Clinicians. 2016;**66**:7-30. DOI: 10.3322/caac.21332

[66] Bray F, Ferlay J, Soerjomataram I, Siegel RL, Torre LA, Jemal A. Global cancer statistics 2018: GLOBOCAN estimates of incidence and mortality worldwide for 36 cancers in 185 countries. CA: A Cancer Journal for Clinicians. 2018;**68**:394-424. DOI: 10.3322/caac.21492

[67] Afshari J, Davoodi S, Samarghandian S. Honey induces apoptosis in renal cell carcinoma. Pharmacognosy Magazine. 2011;**7** 46-52. DOI: 10.4103/0973-1296.75901

[68] Orsolic N, Knezevic A, Sver L, Terzic S, Hackenberger BK, Basic I. Influence of honey bee products on transplantable murine tumours. Veterinary and Comparative. Oncology. 2003;**1**:216-226. DOI: 10.1111/j.1476-5810.2003.00029.x

[69] El-Gendy MMA. In vitro, evaluation of medicinal activity of Egyptian honey from different floral sources as anticancer and antimycotic infective agents. Journal of Microbial and Biochemical Technology. 2010;**2**:118-123. DOI: 10.4172/1948-5948.1000035

[70] Portokalakis I, Hi MY, Ghanotakis D, Nigam P. Manuka honey-induced cytotoxicity against mcf7 breast cancer cells is correlated to total phenol content and antioxidant power. Journal of Advances in Biology and Biotechnology. 2016;**8**:1-10. DOI: 10.9734/jabb/2016/27399

[71] Afrin S, Forbes-Hernandez TY, Gasparini M, Bompadre S, Quiles JL, Sanna G, et al. Strawberry-tree honey induces growth inhibition of human colon cancer cells and increases ros generation: A comparison with manuka honey. International Journal of Molecular Sciences. 2017;**18**:613-632. DOI: 10.3390/ijms18030613

[72] Jaganathan SK, Mandal M. Honey constituents and their apoptotic effect in colon cancer cells. Journal of Apiproduct and Apimedical Science. 2009;**2**:29-36. DOI: 10.3896/IBRA.4.01.2.02

[73] Badolato M, Carullo G, Cione E, Aiello F, Caroleo MC. From the hive: Honey, a novel weapon against cancer. European Journal of Medicinal Chemistry. 2017;**142**:290-299. DOI: 10.1016/j.ejmech.2017.07.064

[74] Tomasin R, Gomes-Marcondes MCC. Oral administration of *Aloe vera* and honey reduces walker tumour growth by decreasing cell proliferation and increasing apoptosis in tumour tissue. Phytotherapy Research. 2010;**25**:619-623. DOI: 10.1002/ptr.3293

[75] Lissoni P, Giani L, Zerbini S, Trabattoni P, Rovelli F. Biotherapy with the pineal immunomodulating hormone melatonin versus melatonin plus *Aloe vera* in untreatable advanced solid neoplasms. Natural Immunity. 1998;**16**:27-33. DOI: 10.1159/000069427

[76] Fauzi AN, Norazmi MN, Yaacob NS. Tualang honey induces apoptosis and disrupts the mitochondrial membrane potential of human breast and cervical cancer cell lines. Food and Chemical Toxicology. 2011;**49**:871-878. DOI: 10.1016/j.fct.2010.12.010

[77] Jubri Z, Narayanan NNN, Karim NA, Ngah WZW. Antiproliferative activity and apoptosis induction by Gelam honey on liver cancer cell line. International Journal of Applied Science and Technology. 2012;**2**:135-141

[78] Kishore RK, Halim AS, Syazana MS, Sirajudeen KN. Tualang honey has higher phenolic content and greater radical scavenging activity compared with other honey sources. Nutrition Research. 2011;**31**:322-325. DOI: 10.1016/j.nutres.2011.03.001

[79] Al-Mamary M, Al-Meeri A, Al-Habori M. Antioxidant activities and total phenolics of different types of honey. Nutrition Research. 2002;**22**:1041-1047. DOI: 10.1016/s0271-5317(02)00406-2

[80] Batumalaie K, Qvist R, Yusof KM, Ismail IS, Sekaran SD. The antioxidant effect of the Malaysian Gelam honey on pancreatic hamster cells cultured under hyperglycemic conditions. Clinical and Experimental Medicine. 2013;**14**:185-195. DOI: 10.1007/s10238-013-0236-7

[81] Centers for Disease Control and Prevention. National Diabetes Statistics Report. Atlanta, GA: Centers for Disease Control and Prevention, US Department of Health and Human Services; 2017

[82] Erejuwa OO, Sulaiman SA, Wahab MS, Sirajudeen KN, Salleh MS, Gurtu S. Antioxidant protection of Malaysian Tualang honey in pancreas of normal and streptozotocin-induced diabetic rats. Annales D'endocrinologie. 2010;**71**:291-296. DOI: 10.1016/j.ando.2010.03.003

[83] Cianciosi D, Forbes-Hernandez TY, Afrin S, Gasparrini M, Reboredo-Rodriguez P, Manna PP, et al. Phenolic compounds in honey and their associated health benefits: A review. Molecules. 2018;**23**:2322-2342. DOI: 10.3390/molecules23092322

[84] Yao LK, Razak SLA, Ismail N, Fai NC. Malaysian Gelam honey reduces oxidative damage and modulates antioxidant enzyme activities in young and middle aged rats. Journal of Medicinal Plants Research. 2011;**5**:5618-5625

[85] Aljadi AM, Kamaruddin MY. Evaluation of the phenolic contents and antioxidant capacities of two Malaysian floral honeys. Food Chemistry. 2004;**85**:513-518. DOI: 10.1016/s0308-8146(02)00596-4

Chapter 5

A Review on Analytical Methods for Honey Classification, Identification and Authentication

Abstract

Authentication of food products is of great concern in the context of food safety and quality. In recent years, interest in honey authenticity in relation with botanical or geographical origin and adulteration has increased. Honey is a ready-to-eat natural food with high nutritional content and gives many health benefits. Authentication of honey has primary importance for both industries and consumers in combatting common honey frauds in the form of mislabeling of honey origin and adulteration with sugar or syrups. Various analytical methods are used for detecting original honey. With a diverse range of equipment and techniques, the conventional analytical methods are still being used in association with advanced techniques as they are part of preliminary screening, processing and product standards. Most of the analytical methods provide indications of pollen distribution, physicochemical parameters and profile analysis of phenolic, flavonoid, carbohydrate, amino acids, aroma and individual marker components. This review provides an overview and summary of instrumental and analytical methods available for honey authentication from conventional to recent molecular techniques. It is useful as a guide to choosing appropriate method for analysis, classification and authentication of honey.

Keywords: adulteration, high fructose sugar syrup, botanical origin, geographical origin, entomological origin

1. Introduction

Honey is a natural sweet substance used by human beings since ancient times. The first written evidence of honey was found in a Sumerian tablet dating back to 2100–2000 B.C. [1]. Honey is defined by the European Union as "the natural sweet substance produced by *Apis mellifera* bees from the nectar of plants or from secretions of living parts of plants or excretions of plant-sucking insects or the living parts of plants, which the bees collect, transform by combining with specific substances of their own, deposit, dehydrate, store, and leave in honeycombs to ripen and mature" [2, 3]. Honey can be classified following its origin, the way it has been harvested and processed. Following honey origin, it is categorised into blossom, honeydew, monofloral and multifloral honeys. Blossom honey is obtained mainly from the nectar of flowers while honeydew or forest honey is produced by bees after they collect "honeydew" from plant saps. Monofloral honey is arising predominantly from a single botanical origin with above 45% of total pollen content from the same plant species, and is named after that plant, such as citrus, manuka and acacia honey [4]. Multifloral honey is also known as polyfloral honey. It has several botanical sources where none is predominant, for example, meadow blossom honey and forest honey. Honey is an important, distinct and widely used food product for nutrient, cosmetic and medicinal purposes.

The main composition of honey is carbohydrates or sugars, which represent 95% of honey dry weight. Honey is a complex mixture of concentrated sugar solution with main ingredients of fructose and glucose. The average ratio of fructose to glucose is 1.2:1 [5]. Sucrose is present in honey at about 1% of its dry weight. The exact proportion of fructose to glucose in any honey depends largely on the source of the nectar. It also contains bioactive compounds like organic acids, proteins, amino acids, minerals, polyphenols, vitamins and aroma compounds [6, 7]. The protein content of honey is normally less than 0.5% with a small fraction of enzymes. The overall quality of honey such as taste, colour and other physical properties are contributed by the non-volatile compounds like sugar, amino acids, minerals and phenolic compounds while aroma of honey is mainly contributed by the volatile components [8]. The compositional criteria prescribed in existing honey directive are requirements relating to concentrations of acidity, apparent reducing sugar which is calculated as invert sugar and apparent sucrose, 5-hydroxymethylfurfural (HMF), mineral content, moisture and water-insoluble solids [9]. HMF is formed from reducing sugars in honey in acidic environment and often used to evaluate honey quality as it is strongly correlated to ageing and overheating of honey [10]. It is set at maximum limit of 40 mg/kg in honey (with a higher limit of 80 mg/kg for tropical honey) by the Codex Alimentarius Standard commission [11].

Honey has various biological properties including antimicrobial, anti-viral, anti-inflammatory, wound and sunburn healing, antioxidant, anti-parasitic, anti-diabetic, anti-mutagenic and anti-tumoral activities [5, 12–17]. Recent pharmacological studies have revealed that natural honeys have potential to reduce risk of gastric and cardiovascular diseases [18] and have beneficial effects on fertility and ameliorating hormones related to fertility [19–21]. With many beneficial properties, honey is highly priced and is also the major target of adulteration.

2. Authenticity of honey

Honey has become the target of adulteration with sugar and/or syrups, for example, cheaper sweeteners from beet or canes like corn syrups (glucose), high fructose corn syrup (HFCS), saccharose syrups, and invert sugar syrups in several countries [22]. In some places, honey is adulterated by bee feeding with sugars or syrups to produce artificial honey. Adulterated honey in the market is often labelled and sold as pure honey and artificial honeys are mislabeled for its botanical or geographical origin [23]. The monofloral honeys are main target for mixing with cheaper multifloral honeys. Monofloral honeys, being the most appreciated by consumers are recognised as better-quality products and they have higher market values [24]. Identification of pure honey and its authenticity have been done based on honey properties. As it becomes an important task for processors, retailers, consumers and regulatory authorities, various analytical methods to measure honey authenticity have been used to detect these honey frauds (**Figure 1**).

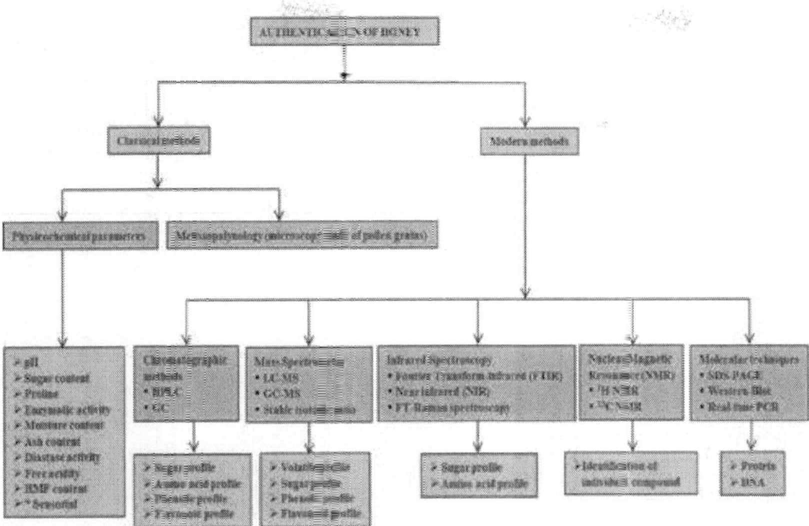

Figure 1.
Conventional and modern analytical methods used for honey authentication.

3. Detection methods and technologies

The classical approach in honey authentication studies is used for determining its botanical origin. Sensory and physicochemical analyses are used in determining monofloral honey origins while the melissopalynological analysis is commonly used to identify floral pollen grains present in honey by microscopic examination [25]. The melissopalynological approach however may not be appropriate for some types of honey like the citrus as the level of pollen content is variable and normally little [26]. The proportion of pollen content is dependent on the plant species, collection season and the nectar yield in male and female flowers. In some cases, pollen can be filtered out in the bee's honey sac and added fraudulently in honey [27]. Due to the significant natural variation of pollen content, this method is now accompanied with sensory analysis and determination of certain physicochemical characteristics. The number of physicochemical parameters necessary for a complete characterisation is very high and melissopalynology has disadvantages of being slow, very tedious to implement and requires a considerable amount of training.

Due to the limitations of the classical authentication techniques, more reliable modern analytical methods are used to determine botanical and geographical origins of honey. The studies include measuring carbohydrate (sugar) profiles [28], mineral content [29], phenolic and flavonoid compositions [30], aroma profile [31, 32] and amino acid composition [33] using advanced analytical tools like chromatographic techniques [28], mass spectrometry (MS)-based techniques [34, 35], vibrational spectroscopy like infrared (IR) and Raman techniques [36], nuclear magnetic resonance (NMR) [37], stable isotope analysis [38, 39] and others such as flame ionisation detectors (FID) or sensor arrays [40, 41].

Several authors studied on the detection of adulterants like exogenous sugars or additions of sugar syrups by evaluating carbohydrate with different analytical techniques [42–44]. Honey is principally constituted by a mixture of different saccharides such as glucose, fructose, tri- and tetrasaccharides while other components are present only in very minimum amount [45]. Botanical classification of honey was previously studied using sugar profile and recently high-performance liquid chromatography (HPLC) and chemometric analysis are used to determine sugar profiles of honey [46, 47]. For evaluating floral origin of honey, its volatile composition is determined using headspace solid-phase microextraction (SPME) and gas chromatography coupled to mass spectrometry (GC–MS) [48]. Researchers have also integrated these techniques with chemometric analysis to classify botanical origin of honeys [48, 49]. In chemometric techniques, principal component analysis (PCA) and linear discriminant analysis (LDA) are used to determine the most influencing variables and similarities in studied honey samples [50, 51]. Recent studies have demonstrated the use of molecular genetics approach in determining composition and geographical origins of honey [24], and entomological origins of honey [52–55]. The advantages and limitations of each analytical technique are reviewed and compared in the following sections.

3.1 Physicochemical parameters for honey identification

Physicochemical parameters such as pH, sugar content, proline, enzymatic activity, moisture content, ash content, diastase activity, free acidity and hydroxymethylfufural (HMF) content could provide useful information of honey origin. Nozal Nalda et al. [46] found significant differences of honey samples in terms of 15 mineral contents (except for iron and zinc) and 8 physicochemical parameters (except for sucrose and HMF) in their large-scale study on 73 different honeys of 7 botanical origins of the ling, heather, rosemary, thyme, honeydew, spike lavender and French lavender honeys. The classification of three monofloral Serbian honeys, the acacia, sunflower and linden could be based on variables such as electrical conductivity (0.10–0.76 mS/cm), free acidity (7.80–42.70 meq/kg) and pH (3.17–5.85) [56]. Silvano et al. [57] reported significant differences in the mean value of HMF, colour, electrical conductivity and sucrose content for honeys harvested from different apiaries such as agricultural, hill and meadow zones of the southeast region of Buenos Aires province in Argentina. **Table 1** presents other comparative studies on physicochemical parameters of different types of honeys by investigators from different regions [49, 50, 58–63]. The physiochemical properties of honey are highly dependent on the type of flowers visited by bees, as well as influenced by seasonal, geographical and climatic conditions. Kek et al. [52, 53] have used physicochemical, antioxidant properties and various chemical profiles which included proximate composition, predominant sugars, HMF content, diastase activity, mineral and heavy metal contents to classify honey by its entomological origin, that is, following the bee speciation of honey bees (*Apis spp.*) or stingless bees (*Heterotrigona spp.*).

The sensory properties variation within honeys due to flora in local habitat can be mirrored by pollen analysis or sensorial studies for honey recognition. Stolzenbach et al. [64] reported that Danish honeys had distinct and unique flavours related to its origin of location. Coupled with sensorial analysis, Castro-Vazquez and co-workers used GC–MS to profile volatile compounds of 49 Spanish honey samples from different botanical origins of citrus, rosemary, eucalyptus, lavender, thyme and heather [65] followed by identifying floral marker origins for lavender and lavandin honeys [66]. Lavandin honey is a monofloral product of recent proliferation obtained from a hybrid of the species *Lavandula angustifolia* and *Lavandula latifolia*. In their study, high concentrations of g-nonalactone, farnesol and acetovanillone, which were identified for the first time as components of honey aroma and lactones, dehydrovomifoliol, 4-methoxyacetophenone and decanal were suggested as chemical markers for authenticating lavandin monofloral honey.

Analytical techniques	Samples	References
Melissopalynological	10 Lavender and 10 lavandin honeys	[66]
Palynological	20 Moroccan honey samples from sunflower, crucifer, carob tree, loeflingia, heather, mint and wood sage	[58]
Sensory properties	205 Slovenian honey from different geographical regions	[50]
	24 multifloral honey samples	[57]
	11 brands of bottled honey from the Indian market	[60]
	21 locally produced Danish honeys	[64]
	49 commercial Spanish monofloral honeys	[65]
	Chesnut honeys	[67]
Physicochemical properties	73 honeys from 7 botanical origins: ling, heather, rosemary, thyme, honeydew, spike lavender and French lavender	[48]* mineral
	30 Uruguayan samples, *Eucalyptus* spp., *Citrus* spp., *Baccharis* spp. and multifloral	[49]* chemical
	205 Slovenian honey from different geographical regions	[50]
	201 samples from 3 unifloral Serbian honeys: acacia, sunflower and linden	[56]
	24 multifloral honey samples	[57]
	20 Moroccan honey samples from sunflower, crucifer, carob tree, loeflingia, heather, mint and wood sage	[58]* colour
	22 Brazilian honey of *Eucalyptus* and *Citrus* spp.	[59]* ash
	11 brands of bottled honey from the Indian market	[60]
	26 honey samples from beekeepers in Lithuania	[61]* carbohydrate & electrical conductivity
	77 honey samples: 53 blossom and 24 suspected honeydew	[62]
	67 samples of Indian honeys	[63]* trace metal
	5 raw Malaysian honey from 4 bee species and 3 commercial honey	[52, 53]* chemical, mineral & antioxidant

**Other properties included in studies besides the general physicochemical properties.*

Table 1.
Physicochemical techniques used for identification of honey.

Similar approach of chemical and sensory characteristics of honey was used to classify chestnut honey geographically [67].

3.2 Chromatographic techniques

Different chromatographic techniques have been reported to determine sugar, amino acids, phenolic and flavonoid profiles of honey samples (**Table 2**). In the early days, Doner et al. [68] determined maltose/isomaltose ratios of honeys and high fructose corn syrup using gas chromatographic (GC) method. They reported that ratios higher than 0.51 indicated adulteration. Kushnir [69] demonstrated a thin layer chromatography separation for oligosaccharides in honey after sample

Sugar profiles	Samples	References
HPLC	2 nectar and honeydew honeys of different geographical and floral origin	[70]
HPLC-PAD	50 honey samples from different regions of Algeria	[28]
HPAEC-PAD	Fir, rosemary, chestnut and thyme honeys	[43]
	17 artisanal and 8 commercial honeys	[72]
HPLC-DAD	160 honey samples from Acacia, jujube, rape, linden, litchi, clover and multifloral	[73]
HPTLC	15 commercial honeys from 3 types of flowers: lime, polyfloral and acacia	[74]
GC–MS/FID and LC- PAD	280 French honeys from 7 monofloral varieties: 50 acacia, 38 chestnut, 28 rape, 53 lavender, 37 fir, 38 linden and 36 sunflower	[75]
Amino acid profiles		
GC	45 honey samples from the UK, Australia, Argentina and Canada	[76]
HPLC	A variety of different honey samples	[77]
	280 French honeys from 7 monofloral varieties: 50 acacia, 38 chestnut, 28 rape, 53 lavender, 37 fir, 38 linden and 36 sunflower	[78]
	7 different floral types of Serbian honey: acacia, linden, sunflower, rape, basil, giant goldenrod and buckwheat	[80]
LC-ECD	29 honeys: 12 of floral origin and 17 from honeydew	[79]
	10 acacia honeys and 10 rape honeys	[81]
Phenolic and/or flavonoid profiles		
HPLC	9 monofloral eucalyptus honeys from Australia	[82, 84]
	Australian Melaleuca, Guioa, Lophostemon, Banksia and Helianthus honeys	[8, 83]
	Polish honey from heather and buckwheat	[85]
	119 unifloral honeys from 14 different geographical regions	[87]
	7 honey samples: acacia, sulla, thistle and citrus honeys	[88]
	4 types of Spanish honey: floral origin of citrus, rosemary and polyfloral and forest origin of honeydew	[92]
	3 Malaysian tropical honeys: Tualang, Gelam and Borneo and 1 Manuka honey	[93]
HPLC-DAD	40 samples of Robinia honey from Croatia	[30]
	38 sage honey samples from Croatia	[89]
	5 types of honey: 2 milk vetch, 1 wild chrysanthemum, 1 jujube flower and 1 acacia	[94]
HPLC-ECD	Chinese citrus honey	[90]
HPLC-UV	90 Italian honeys	[86]
HPLC-UV and GC–MS	Lemon blossom honey and orange blossom honey	[91]

Table 2.
Chromatography-based techniques used for honey authentication.

clean-up by a charcoal-Celite column. Later, Lipp et al. [70] developed a medium-pressure liquid chromatography method using a charcoal/celite mixture for detection of conventional and high fructose syrup in honey at a level as low as 1% of the total mixture. Honey adulterated with commercial sweeteners could be identified easily based on the sucrose level which was slightly higher than the sucrose content of natural honey [71]. Ouchemoukh et al. [28] investigated sugar profiles of 50 honey samples from different regions of Algeria by HPLC with pulsed amperometric detection (PAD) where 11 sugars were quantified. The average values of fructose and glucose were in the range of 35.39–42.57% and 24.63–35.06%, respectively. The sucrose, maltose, isomaltose, turanose and erlose were found in nearly all the analysed samples, while raffinose and melezitose were present in few samples. Cordella et al. [43] used high performance anion exchange chromatographic method with pulsed amperometric detection (HPAEC-PAD) to detect adulterated honey samples with industrial bee-feeding sugar syrups. Morales et al. [72] determined high molecular weight oligosaccharides of 9 sugar syrups and 25 honey samples using HPAEC-PAD for detection of honey adulterated with corn syrups and high fructose corn syrup (HFCS). Xue et al. [73] developed an HPLC-diode-array detection (DAD) method to detect honey adulteration using rice syrup and identified the adulterant compound from rice syrup as 2-acetylfuran-3-glucopyranoside. Puscas et al. [74] developed a simple and economical analytical method for detecting adulteration of some Romanian honeys based on high-performance thin-layer chromatography (HPTLC) combined with image analysis. The method was then applied for quantitative analysis of glucose, fructose and sucrose contents from different types of commercially available Romanian honeys. Cotte et al. [75] have developed a method using LC coupled with a pulsed amperometric detector (PAD) to assay fructose and glucose, and GC coupled with a flame ionisation detector (FID) to determine the entire profile of di- and trisaccharides.

Gilbert et al. [76] determined 17 free amino acids in 45 honey samples collected from the UK, Australia, Argentina and Canada by using GC. Pawlowska and Armstrong [77] used HPLC methods to determine concentrations of proline, leucine and phenylalanine and their enantiomeric ratios from various honey samples. Cotte et al. [78] claimed to be the first to use HPLC system to discriminate different botanical origins of honey using acid amino analysis. They were successful in characterising Lavender honey. Iglesias et al. [79] developed a reliable and simple method of liquid chromatography–electrochemical detection (LC-ECD) to detect adulteration of acacia honey which was added with rape honey at different levels. Fingerprints of authentic honeys showed that contents of chlorogenic acid were higher in acacia honey, while those of ellagic acid were much lower in rape honey. The free amino acids profile of seven different floral types of Serbian honey (acacia, linden, sunflower, rape, basil, giant goldenrod and buckwheat) from 6 different regions were analysed to discriminate honeys by their botanical origins [80]. Wang et al. [81] suggested chlorogenic acid and ellagic acid as possible markers of acacia and rape honeys authenticity study using LC-ECD.

Currently, research is predominantly focused on determination of phenolic and flavonoid profiles due to their pharmacological properties. Many studies have reported phenolic and flavonoid profiles of honey in relation to botanical and geographical origins using various chromatography techniques [8, 82–87]. The determination of phenolic compounds in honey includes removal of matrix components (especially sugars) and analysis of pre-concentration of analytes using HPLC techniques [88]. Yao et al. [82, 83] analysed phenolic acids and flavonoids along with two abscisic isomers related to botanical origins of nine monofloral Eucalyptus honeys and other five botanical species (Melaleuca, Guioa, Lophostemon, Banksia and Helianthus) from Australia. In Croatia, flavonoid profiles of Robinia honeys [30] from two production seasons and monofloral sage honey [89] were measured using HPLC/DAD method. The respective honeys showed a common and specific flavonoid profile, but the contents varied between seasons. Liang et al. [90]

established a sensitive and accurate method using HPLC-ECD for simultaneous separation and determination of four phenolic compounds, including caffeic acid, p-coumaric acid, ferulic acid and hesperetin in Chinese citrus honey which were 6–14 times greater than those obtained with DAD. Escriche et al. [91] evaluated flavonoids (naringenin, hesperetin, chrysin, galangin, kaempferol, luteolin, pinocembrin and quercetin) and phenolic acids (caffeic acid and p-coumaric acid) together with 37 volatile compounds in the differentiation between lemon blossom honey (*Citrus limon*) and orange blossom honey (*Citrus spp.*). Naringenin and caffeic acid were the main components in all samples. They have concluded that botanical origin affected the profile of flavonoids and phenolic compounds sufficiently to permit discrimination of honeys, that is, hesperetin in citrus honey, kaempferol, chrysin, pinocembrin, caffeic acid and naringenin in rosemary honey, and myricetin, quercetin, galangin and particularly p-coumaric acid in honeydew honey [92]. Khalil et al. [93] investigated phenolic acid and flavonoid contents of Malaysian Tualang, Gelam and Borneo tropical honeys in comparison with Manuka honey. A total of six phenolic acids (gallic, syringic, benzoic, trans-cinnamic, p-coumaric and caffeic acids) and five flavonoids (catechin, kaempferol, naringenin, luteolin and apigenin) were identified. Jasicka-Misiak et al. [85] determined phenolic profiles of Polish honey samples from heather (*Calluna vulgaris* L.) and buckwheat (*Fagopyrum esculentum* L.). The results revealed that the samples of the same unifloral honeys registered a similar qualitative but slightly quantitatively different phenolic characteristic profiles. Perna et al. [86] identified and quantified phenolic acids, flavonoids and vitamin C in 90 Italian honeys of different botanical origins (chestnut, sulla, eucalyptus, citrus and multifloral) using HPLC–UV analysis. The results showed a similar but quantitatively different phenolic profile of the studied honeys. Zhang et al. [94] described the use of second-order calibration for development of HPLC-DAD method to quantify nine polyphenols in five types of honey samples. Greek unifloral honeys (pine, thyme, fir and orange blossom) were analysed and classified according to botanical origin based on phenolic content (quercetin, myricetin, kaempferol, chrysin and syringic acid) by HPLC analysis [87]. Recently, Campone et al. [88] described a novel approach for a rapid analysis of 5 phenolic acids and 10 flavonoids in honey using dispersive liquid-liquid microextraction followed by HPLC analysis. The proposed new method, compared with commonly used method in analysis of phenolic compounds in honey, provided similar or higher extraction efficiency with exception of the most hydrophilic phenolic acids. These chromatographic techniques provide quite complex chromatograms and coupled with appropriate analyses, can classify honey according to their botanical, geographical and entomological origins.

3.3 Mass spectrometry integration with chromatography techniques

The LC–MS and GC–MS methods are used to separate and identify semi-volatile and volatile components in honey. Volatiles influence chiefly to honey flavour and to its variation of floral origin. **Table 3** shows various mass spectrometry techniques used for detection of honey origin. The headspace (HS) solid-phase microextraction (SPME) is the most preferred technique for determining concentration of honey volatiles [91, 95, 96]. With the HS-SPME followed by GC/MC analysis, various volatile components in honey have been detected, identified and quantified, for example, 35 volatile components from Spanish honeys [97], 62 compounds from 28 Greek honeys [98], 31 compounds from 16 samples from mostly European countries [99] and 26 commonly available volatiles in 70 authentic Turkish honey from 9 different floral types [100]. Bianchi et al. [101] developed a HS-SPME method and characterised 40 volatile compounds of Italian thistle honey. In Spain, Soria et al. [102]

Volatile Profiles	Samples	References
GC–MS	80 raw unifloral honey: acacia, sunflower and tilia or lime from Spain, Romania and Czech Republic	[104]
	70 Turkish honey from 5 different geographical regions	[100]*LC/MS for free amino acids
SPME/GC–MS	36 samples of orange honey	[35]*chiral volatile compounds
	42 unifloral honey samples from 5 floral origins: alfafa, sunflower, white clover, carob and calden	[48]*chemometrics
	28 thyme honeys samples from different locations all over Greece	[98]
	Commercial types of European honey	[99]
	46 artisanal honey samples collected in Madrid province, Central Spain	[102]
	40 commercial honey samples	[103]
	Rapeseed, chestnut, orange, acacia, sunflower and linden honeys	[106]*chiral volatile compounds
HS-SPME/GC/MS	7 samples of thistle honey	[101]
	77 unifloral honey samples of different botanical origins	[105] *fingerprinting
HS-SPME/ GC × GC-TOF-MS	374 honey samples of Corsican, non-Corsican–French, Italian, Austrian, Irish and German	[31, 34]
SIFT-MS	9 New Zealand honeys: beech honeydew, clover, kamahi, manuka, rata, rewarewa, tawari, thyme and vipers bugloss	[109]
	10 Ohio and Indiana honey samples: star thistle, blueberry, clover, cranberry and wildflower	[110]
SPME and LC-DAD-ESI/MS	Acacia, linden, chestnut, fir, spruce, floral and forest honey from Slovenia	[51]
HPLC-DAD-MS/MS	Strawberry tree honeys	[111]
	187 honey samples: 98 chaste honey and 89 rape honey	[112]
SPME and UPLC-PDA-MS/MS	Manuka, honeydew, heather, chestnut, and eucalyptus honeys from various geographical origins and ages	[113]
UPLC-Q/TOF-MS	Sunflower, lime, clover, rape and honeydew honey	[114]
Sugar Profile (Carbohydrates and Syrups)		
GC–MS	Honeys from avocado, strawberry tree, willow, loquat, almond tree, fir, evergreen oak, *Anthyllis cytisoides*, *Satureja montana*, Teide broom, agave and tajinaste	[45]
	20 honey samples: 16 nectar and 4 honeydew honeys	[107]
	107 floral honeys: heather, rosemary, eucalyptus and citrus	[108]

**Additional feature of measurements.*

Table 3.
Mass spectrometry-based techniques used for honey authentication.

differentiated honey origins of mountain and plain by characterising their volatile compositions using 46 artisanal honey samples from different places of Madrid province. Later in 2019, the same co-workers characterised Spanish honeys to their botanical origin using 132 volatile compounds from 40 honey samples [103].

Although volatile compounds can be used to differentiate honey according to country, Juan-Borrás et al. [104] concluded that botanical origin differentiates honey samples better than geographical origin in their research using acacia, sunflower and tilia honeys from Spain, Romania, and Czech Republic. In extension to HS-SPME/GC–MS, Baroni et al. [48] used chemometric to determine organic volatile compound pattern to characterise 42 unifloral honey samples from 5 floral origins while Aliferis et al. [105] used finger-printing to discriminate and classify Greek honeys by their plants and geographical origins. Špánik et al. [106] and Verzera et al. [35] developed a new analytical approach based on enantiomeric ratio investigation of chiral volatile constituents using SPME-GC–MS to evaluate authenticity of honeys.

Besides volatile components in honey, GC–MS has also been used to identify sugar compounds in honey. de la Fuente et al. [45] identified various carbohydrate markers in identifying botanical origins of Spanish honey such as disaccharide maltulose and carbohydrate alcohols of perseitol in avocado honeys and melezitose and quercitol in evergreen oak honeys. Carbohydrate composition of 20 honey samples (16 nectar and 4 honeydew honeys) and 6 syrups has been studied by GC and GC–MS in order to detect differences between both sample groups. The presence of difructose anhydrides (DFAs) in these syrups was described for the first time [107]. Ruiz-Matute et al. [108] developed a GC–MS method for the detection of honey adulteration with high fructose inulin syrups. Inulotriose proved to be the best marker of honey adulteration with these syrups since it was not detected in any of the analysed honey samples.

Beyond HS-SPME, Cajka et al. [31] and Stanimirova et al. [34] investigated the use of the system with comprehensive two-dimensional gas chromatography-time-of-flight mass spectrometry (GC × GC-TOFMS) to analyse volatile components in honey samples. Langford et al. [109] and Agila and Barringer [110] used selected ion flow tube-mass spectrometry (SIFT-MS), a growing technology that quantifies volatile organic compounds at low concentrations (usually parts-per-trillion, ppt levels) to determine aromas arise from volatile organic compounds in headspace of different monofloral honeys from New Zealand and Ohio, and Indiana, respectively. Furfural, 1-octen-3-ol, butanoic and pentanoic acids were the volatiles with the highest discriminating power among the different types of floral honey [110]. Bertoncelj et al. [51] used diode array detection system and electrospray ionisation mass spectrometry (LC-DAD-ESI/MS) to analyse flavonoid profiles of seven types of Slovenian honey upon solid-phase extraction followed by liquid chromatography to study their botanical origins. The performance liquid chromatography–diode array detection–tandem mass spectrometry (HPLC–DAD–MS/MS) method was used to trace floral origin of strawberry tree honey [111] and chaste honey and rape honeys [112]. Kaempferol, morin and ferulic acid were used as floral markers to distinguish chaste honey from rape honey. Oelschlaegel et al. [113] used photodiode array detection (PDA) with ultra-performance liquid chromatography (UPLC-PDA-MS/MS) to analyse volatile composition of numerous Manuka honeys after solid-phase extraction and identified kojic acid, unedone, 5-methyl-3-furancarboxylic acid, 3-hydroxy-1-(2-methoxyphenyl)penta-1,4-dione and lumichrome in Manuka honey for the first time. Other technologies used include the system of Ultra Performance Liquid Chromatography-Quadrupole/Time of flight-mass spectrometry (UPLC-Q/TOF-MS) where it was possible to identify several components which cannot be detected by diode array using combination of detection with retention time for accurate molecular mass to obtain phenolic acids and flavonoids from ethyl acetate extracts of different honeys (sunflower, lime, clover, rape and honeydew) [114].

3.4 Stable isotopic ratio mass spectrometry (IRMS)

The stable isotope ratio analysis using mass spectrometry may be used to detect adulterated honey samples based on the principle of different $\delta^{13}C$ or 13C/12C ratio [115–117]. Honey-producing plants, as well as sugar beets belong to the C3 plants while sugar cane, corn and other major source of adulterating syrups are from the C4 plants. Their different photosynthetic pathways result in a different metabolic enrichment of the 13C isotope. The slower reacting $^{13}CO_2$ is depleted to a larger extend in C3 plants than in C4 plants during the CO_2 fixation (kinetic isotope effect). Thus, it is possible to detect the addition of cheap C4 sugar because of its different $\delta^{13}C$ value ranging from −22 to −33δ‰ for honey from C3 plants, −10 to −20δ‰ for honey from C4 plants and −11 to −13.5δ‰ in honey from Crassulacean Acid Metabolism plants (pineapple and cactus). When C4 sugar is added to pure honey, the $\delta^{13}C$ value of honey will be altered where honey with $\delta^{13}C$ values less negative than −23.5‰ is suspected to be adulterated. The protein extracted from honey can be used as an internal standard for the determination of adulteration in honey as the corresponding $\delta^{13}C$ of protein extract will remain constant. The difference in $\delta^{13}C$ between honey and its associated protein extract accepted is −1δ‰ deviation at least, which provides the international benchmark of 7% of C4 sugar added [116]. **Table 4** lists honey authentication studies based on IRMS analysis.

Isotope $\delta^{13}C$	Samples	Reference
IRMS	49 samples of honey	[115]
	40 samples of honey of various botanical origins produced in Brazil, and 8 imported samples, 1 from Argentina, 3 from the USA and 4 from Canada	[116]
	73 Italian honey samples from 6 varieties: chestnut, eucalyptus, heather, sulla, honeydew and wildflower	[117]
	100 pine honey samples	[118]
	13 different brands of honey samples	[120]
	271 Slovenian honey samples from 7 floral types and 4 geographical regions	[38]* $\delta^{15}N$
EA-IRMS	140 honeys from 7 different botanical origins: acacia, chestnut, rapeseed, lavender, fir, linden and sunflower	[121]
	31 Turkish honey from different sources and regions: flower, pine and chestnut and 43 commercial honey	[123]
	58 honey samples: Northeast China black bee, chaste, acacia, clover, chaste, flowers and jujube honeys	[124]
	Commercial honey	[22]
	516 authentic honeys from 20 European countries	[39]* $\delta^{2}H$, $\delta^{15}N$, $\delta^{34}S$
HPLC-IRMS	79 commercial honey samples	[119]
EA/LC-IRMS	451 authentic honeys	[122]
IRMS and SNIF-NMR	102 French honey from 97 varieties: acacia, chestnut, rape, lavender, fir, linden and multifloral	[44]
Flow isotope IRMS	Manuka honey from New Zealand	[125]

**Include other isotopes measurements.*

Table 4.
Isotope-based techniques to determine botanical and geographical origins of honey.

Most honey adulteration studies using carbon isotope ratio mass spectrometry (IRMS) are tested on the addition of C4 plant sugars such as HFCS in Turkish pine honey [118], beet sugar products or corn syrup addition in Spanish honey [115] and HFCS, glucose syrup from corn starch and saccharose syrups from beet sugar [22]. Cabanero et al. [119] determined individual sugars like sucrose, glucose and fructose ^{13}C isotope ratios from honeys of various botanical and geographical origins sources which have been adulterated with beet sugar (C3) and/or C4 sugars like cane sugar, cane syrup, isoglucose syrup and high-fructose corn syrup (HFCS). They developed the first isotopic method that allows beet sugar addition detection. Cengiz et al. [120] provided further validation parameters, such as the limit of detection, limit of quantification and recovery for honey adulteration detection method developed using IRMS.

Besides carbon isotope ratios for sugar and protein, Kropf et al. [38] used the nitrogen stable isotope in honey authentication study using 271 honey samples from 4 different geographical regions of Slovenia while Schellenberg et al. [39] used multiple element stable isotope ratios of hydrogen, carbon, nitrogen and sulphur stable for 516 authentic honeys from 20 European regions. Daniele et al. [121] developed a method to discriminate honeys from seven botanical origins, based on organic acid analysis. The authors suggested that by combining various organic acid contents and values of isotopic ratio through statistical processing by PCA, it is possible to discriminate honey samples as a function of their botanical origin.

The IRMS system used are also enhanced with an elemental analyser and to a liquid chromatograph [122–124] in studies to determine adulteration of C4 plant sugar content in honey. Cotte et al. [44] used site-specific natural isotopic fractionation determined by NMR to first determine their potentials for characterising the substance and then to detect adulteration. They found that the system is limited by detection of a syrup spike starting only at 20%. Frew et al. [125] used a continuous flow IRMS in their adapted method for removing pollen in Manuka honey to improve authenticity detection.

3.5 Infrared spectroscopy

Fourier Transform-infrared (FTIR) spectroscopy, near infrared (NIR) and FT-Raman spectroscopy methods have also proved their great potential in food authentication studies. In recent years, these spectroscopic techniques have achieved wide acceptance in the field of food sciences for quantitative and qualitative analysis because of its advantages of collecting high-spectral-resolution data over a wide spectral range. They have been applied successfully in honey authentication studies from the aspects of identification of honey origins and determination of adulterants (**Table 5**).

Ruoff and co-workers have used chemometric evaluations of the spectra measured in various European honeys using near-infrared spectroscopy [126] and front-face fluorescence spectroscopy [127] to authenticate botanical and geographical origins of honey. Important physical and chemical measurands in honey such as sucrose and fructose/glucose ratio using near-infrared spectrometry [128] and by mid-infrared spectrometry (MIR) [129] were also helpful for assessing honey adulteration. Dvash et al. [130] reported the application of NIR reflectance spectroscopy to determine concentration of perseitol, a sugar that is specific to avocado honey. Qiu et al. [131] studied influences of various sample presentation methods and regression models and presented that the spectroscopic technique could accurately determine moisture, fructose, glucose, sucrose and maltose contents in honey samples. Woodcock and co-workers used chemometric tools in NIR spectroscopic studies to differentiate different geographical origins of honey which included Corsican

Technique	Samples	References
NIR	74 commercial honey	[131]
	Avocado honey	[130]
	167 unfiltered honey samples and 125 filtered honey samples	[133]
	Honey from 6 floral origins	[134]
	30 honey samples from Galicia	[135]
	75 honey samples from Ireland	[138]
	68 honey samples from 6 floral origins: *Brassica* spp., *Zizyphus* spp., *Citrus* sp., *Robinia pseudoacacia*, *Vitex negundo var. heterophylla* and multiflora	[139]
	144 honey samples: 70 unadulterated honey samples from different apiary of Beijing and 74 honey samples from local grocery stores	[140]*fibre optic diffuse reflectance
	Authentic, unfiltered honeys of Corsican and non-Corsican	[132]* spectral fingerprinting
FTIR	37 honey samples from different regions in the world	[36]
	1075 honey samples from Germany	[136]
FT-NIR	364 samples from 7 unifloral and polyfloral honeys	[126]
	421 honey samples	[128]
FTMIR	144 honey samples from 7 different crops, polyfloral and honeydew honeys	[129]
	Commercial honey, orange blossom, clover and buckwheat	[137]
Front-Face Fluorescence Spectroscopy	371 samples from 10 unifloral and polyfloral honeys	[127]
FT-Raman	Honey samples of clover, orange and buckwheat	[142]
	Corsican samples from other geographical origins: France, Italy, Austria, Germany and Ireland	[143]
i-Raman	74 authentic honey samples from 10 floral origins	[141]
FTMIR-ATR	99 authentic honey samples from artisanal beekeepers in Ireland	[146]
	580 samples of artisanal Irish honeys	[147]
FTIR-ATR	150 honey samples from Europe and South America	[148]
	Honey samples from 4 different states of Mexico	[149]

**Additional features of measurements.*

Table 5.
Infrared spectroscopy-based techniques used in honey authentication.

and non-Corsican [132], and honeys from Ireland, Mexico, Spain, Argentina, Czech and Hungary [133]. Using NIR, Liang et al. [134] has successfully classified 147 honey samples from six floral origins at perfect classification rate of 100% and Latorre et al. [135] differentiated Galacian honey by protected "Mel de Galicia" geographical location. Using FTIR, Etzold and Lichtenberg-Kraag [136] determined various botanical origins of honey based on physical–chemical characterisation while Wang et al. [36] determined geographical origins based on sugar profiles. For honey adulterants like commercial syrup including glucose and fructose, Sivakesava and Irudayaraj [137] classified simple and complex sugar adulterants using mid-NIR, Downey et al. [138] and Zhu et al. [139] have reported the use of transflectance spectra while Chen et al. [140] used a fibre optic diffuse reflectance probe in their NIR systems.

The other different spectroscopy technique in material characterisation is known as the FT-Raman spectroscopy. It measures relative frequencies at which a sample scatters radiation unlike the IR spectroscopy, where it measures absolute frequencies at which a sample absorbs radiation. Li et al. [141] studied the potential of Raman spectroscopy for detecting fructose corn syrup and maltose syrup adulterants in honey. Paradkar and Irudayaraj [142] investigated on adulterants like cane and beet invert in honey. Pierna et al. [143] used FT-Raman spectroscopy to differentiate Corsican honeys from other regions in France, Italy, Austria, Germany and Ireland.

Despite being relevant in honey authentications, IR or Raman spectroscopic techniques are known to give problems to food samples during analysis as prolonged exposure of food sample to laser beam may lead to sample destruction due to heat. The water absorption is very intense in the mid-IR region. A shorter irradiation time and increasing number of scans are recommended to avoid this problem [144]. de la Mata et al. [145] performed attenuated total reflection (ATR) to overcome this problem. FTIR spectroscopy and attenuated total reflection (ATR) sampling technique were used to study botanical origin of honey sample at mid-infrared spectra [146–147]. Hennessy et al. [148] used the similar but with a germanium ATR to verify origin of honey samples from Europe and South America. Gallardo-Velázquez et al. [149] quantified three different adulterants, corn syrup, HFCS and inverted sugar (IS) in honeys of four different locations in Mexico.

3.6 Nuclear magnetic resonance

Nuclear magnetic resonance (NMR) spectroscopy provides important structural information for a wide variety of food components and is recognised as one of the main analytical techniques for authentication of food products as it is strongly focused on both structural and chemical characterisation [150, 151]. **Table 6** lists researches which have used NMR technique to authenticate honey samples in relation to geographical or botanical origins and honey adulteration. The more common NMR experiments are the one-dimensional (1D) referred as NMR at ^{1}H or ^{13}C spectra and the two-dimensional (2D) NMR referred as the classical $^{1}H^{13}C$ heteronuclear multiple-bond correlation (HMBC). One-dimensional ^{1}H NMR spectra was used to profile saccharides of honey from different countries [152, 153]. Boffo et al. [154] discriminated botanical origins of Brazlian honeys from the eucalyptus, citrus and wildflower origins. Donarski et al. [37] used cryoprobe ^{1}H NMR spectroscopy for verification of Corsican honey's geographical locations in Europe and later used biomarkers to identify botanical origins of sweet chestnut and strawberry-tree honeys [155]. Zielinski et al. [3] proposed phenylacetic acid and dehydrovomifoliol as markers of Polish heather honey, confirmed 4-(1-hydroxy-1-methylethyl)cyclohexane-1,3-dienecarboxylic acid as a marker of lime honey and reported that formic acid and tyrosine as the most common characteristic compounds of buckwheat honey. Using signals of protons and carbon of the methylene group of quercitol in ^{1}H and ^{13}C NMR spectra of honey, Simova et al. [156] identified and discriminated oak honeydew honey from all other honey types of honeydew honeys. Quercitol is considered as a good botanical marker for the genus Quercus which the oak tree belongs. Beretta et al. [157] used ^{1}H NMR profile coupled with electrospray ionisation-mass spectrometry (ESI-MS) and two-dimensional NMR analyses to seek reliable markers of the botanical origin of Italian

Spectrum	Samples	References
^{1}H NMR	35 honey samples from multifloral, heather, lime, rape, buckwheat and acacia	[3]
	180 Corsican honey samples	[37]
	57 samples from different countries	[153]
	46 honey samples from flowers of citrus, eucalyptus, assa-peixe and wildflowers	[154]
	374 samples from Austria, France, Germany, Ireland and Italy	[155]
	118 honey samples of 4 different botanical origins: chestnut, acacia, linden and polyfloral	[159] *chloroform extracts
	353 honey samples from acacia, chestnut, linden, orange, eucalyptus, honeydew and polyfloral	[160] *chloroform extracts
^{1}H NMR and ^{13}C NMR	23 samples of polyfloral and 18 samples of acacia honey	[152]
	24 honey samples of oak and others	[156]
^{1}H NMR and ^{1}H-^{13}C HMBC	63 authentic and 53 adulterated honey samples	[42]
	71 honey samples: robinia, chestnut, citrus, eucalyptus and polyfloral	[158]
^{1}H NMR-ESI-MS and ^{1}H-^{13}C HMBC	44 commercial Italian honeys from 20 different botanical sources	[157]
LF^{1}H NMR	Pure blossom honey samples	[162]
	80 samples from eucalyptus, orange, Barbados cherry, cashew tree, assa-peixe, assa-lipto and Cipo-Uva	[161]

**Additional features of measurements.*

Table 6.
Nuclear magnetic resonance (NMR) techniques in honey authentication.

honeys. Lolli et al. [158] used both ^{1}H NMR and heteronuclear multiple bond correlation spectroscopy (HMBC) to characterise five different floral sources of Italian honey. ^{1}H NMR spectra of chloroform extracts was developed and used to study non-volatile organic honey components for botanical origin characterisation of chestnut, acacia, linden and polyfloral honeys where specific markers were identified for each of the six monofloral Italian honeys [159, 160]. Bertelli et al. [42] investigated adulterated honey falsified by intentional addition of different concentrations of commercial sugar syrups using one-dimensional (1D) and two-dimensional (2D) NMR coupled with multivariate statistical analysis. Ribeiro and co-workers used low field nuclear magnetic resonance spectroscopy (LF ^{1}H NMR) to classify Brazilian honeys into eight different botanical sources [161] and to differentiate honey adulterated by HFCS in different concentrations from 0% (pure honey) to 100% (pure high fructose corn syrup) [162].

3.7 Molecular techniques

Honey contains only about 0.2% protein [163] and it originates from bee and nectar of plants [164–165]. Honey proteins appear in the form of enzymes, predominantly diastase (amylase), invertase and glucose oxidase. Others, including catalase and acid phosphatase, can also be present, depending on the type of floral source and recently proteolytic enzymes have been described in honey. The major proteins in honey possess different molecular weights depending on its bee species. Thus, protein- and DNA-based honey authentication methods such as the sodium dodecyl sulfate polyacrylamide gel electrophoresis (SDS–PAGE) and real-time PCR are used to identify, authenticate and classify honey samples (**Table 7**).

Marshall and Williams [166] used SDS-PAGE, the high-resolution two-dimensional electrophoresis method with methylamine-incorporating silver stain to characterise trace proteins of Australian honeys. Their study revealed that honey protein constituents are predominantly of bee origin. Lee et al. [167] used

Protein-based methods	**Samples**	**References**
Western-blot	Honeys from *Prosopis caldenia*, *Prosopis* sp., *Eucalyptus* sp., *Helianthus annuus*, *Melilotus albus* and *Larrea divaricata*	[168]
	Honey samples from beekeepers and markets throughout Korea	[169]
	RJ proteins from Slovakia	[170]
ELISA	RJ protein Apalbumin 1 in honey samples from acacia, linden, rapeseed, dandelion and chestnut	[171]
SDS-PAGE	Native and foreign bee-honeys	[167]
	Honeys from *Prosopis caldenia*, *Prosopis* sp., *Eucalyptus* sp., *Helianthus annuus*, *Melilotus albus* and *Larrea divaricata*	[168]
	Honey samples from beekeepers and markets throughout Korea	[169]
	Australian honey samples	[166]* high-resolution 2-dimensional electrophoresis
DNA-based methods		
Manual tracking plant, fungal, and bacterial DNA	Commercial, eucalyptus and lemon honeys	[174]
Manual, QIAQuick PCR Purification Kit	Honeydew honey with multifloral, wild flower and rape honeys, *Acacia*-with multifloral honey	[177]
CTAB	1 pine honey, 2 wild honey, 5 polyfloral honey	[175]* pollen
	3 different apiaries	[176]
Other commercial kits methods besides CTAB		
NucleoSpin Plant, Wizard methods and DNeasy Plant Mini Kit	*Calluna vulgaris*, *Lavandula* spp., *Eucalyptus* spp. and a multifloral honey	[24]
NucleoSpin Isolation Food Kit, Wizard Magnetic DNA Purification and DNeasy Mericon Food Kit	14 types of raw honey from *Apis dorsata*, *Apis mellifera*, *Apis cerana* and *Heterotrigona itama*	[54, 55]
DNeasy Tissue Kit	One regional origin (Pyrenean honey) and one worldwide mix of different honeys (wild flower honey)	[172]

RJ, Royal jelly.
**Additional features of measurements.*

Table 7.
Molecular techniques for honey authentication.

SDS-PAGE to differentiate native bee-honey (NBH) and foreign bee-honey (FBH) from different molecular weight found in major protein of NBH at 56 kDa and FBH at 59 kDa which were used as protein markers to differentiate NBH and FBH. Baroni et al. [168] reported on the development of a novel method based on honey proteins to determine floral origin of honey samples using SDS-PAGE immunoblot or Western-Blot techniques. The Western-Blot is done to confirm the presence, absence and expression level of a protein of interest using specific antibodies while SDS-PAGE separates protein based on molecular weight. Won et al. [169] distinguished honey produced by two different bee species, *Apis mellifera* and *Apis cerana* by the difference in molecular weight of their major proteins (56 and 59 kDa) using SDS–PAGE and later used the purified proteins as antigens for antibody reactions in rats. The Western-blot method verified differences in major proteins' surface structure thus can be used to differentiate the different honey bee species. Besides honey, royal jelly, another product secreted by honeybee workers as food to be fed to the larvae which will be raised as the potential queen bee, is known to have high protein contents. Simuth et al. [170] reported the presence of royal jelly (RJ) proteins in honey collected from nectars of different plants, origin and regions and in honeybee's pollen by Western-blot analysis using polyclonal antibodies raised against water-soluble RJ-proteins. They authors suggested that the Apalbumin-1 is a common component of honeybee products and thus is an appropriate marker tool for detecting adulteration of honey by means of immunochemical methods. Bilikova and Simuth [171] developed the 55 kDa major protein of royal jelly, named apalbumin 1 (an authentic protein of honey and pollen pellet), and quantified it by an enzyme-linked immunosorbent assay (ELISA) using specific polyclonal anti-apalbumin 1 antibody.

In recent food authentication methods, DNA-based method is used and regarded as the most reliable, rapid and reproducible technique to detect adulteration and origin of food materials (**Table 7**). Valentini et al. [172] proposed a new method for investigating plant diversity and geographical origin of honey using a DNA barcoding approach that combines universal primers and massive parallel pyrosequencing. Laube et al. [173] developed a DNA-based method for characterisation of plant species in honey which further used as geographical origin indication. They identified PCR markers for detection of plant species related to "Miel de Corse", a protected designation of origin honey and "Miel de Galicia", a protected geographical indication (PGI) honey as well as German and English honeys. Soares et al. [24] used DNA-based methods to identify botanical species of honey. They used five DNA extraction methods combined with three different sample pre-treatments on four honey samples, three monofloral and one multifloral. The different DNA extraction procedures were compared in terms of DNA integrity, yield, purity and amplification targeting universal and ADH1 specific genes of *C. vulgaris* where they identified monofloral heather honey successfully. Although these molecular techniques give appreciable results, a prior knowledge about the plant species is required to identify origin of honey samples. Besides detecting plants, Olivieri et al. [174] used DNA-based methods to detect fungi and bacteria in honey due to potential risks evoked by microorganisms, allergens or genetically modified organisms. Guertler et al. [175] developed an automated DNA extraction method from pollen in honey. The authors altered several components and extraction parameters and compared the optimised method with a manual CTAB buffer-based DNA isolation method. The automated DNA extraction was faster and resulted in higher DNA yield and sufficient DNA purity. The results obtained from real-time PCR after automated DNA extraction are also comparable to that of manual DNA extraction procedure. Jain et al. [176] introduced a protocol for DNA extraction from honey using modified CTAB-based protocol. Waiblinger et al. [177] described an in-house and interlaboratory validation of a DNA extraction method from pollen in a unifloral rape honey with several multifloral honeys. While most DNA studies on honey have focused on botanical species identification, Kek et al. [54, 55] have determined the best DNA extraction of bees in honey and introduced entomological identification of honey based on bee mitochondrial 16S rRNA and COI gene sequences.

3.8 Other techniques

Table 8 presents other methods which included instrumental and improvised analytical procedures used for honey identification and authentication. Cordella et al. [178–179] used differential scanning calorimetry (DSC) to study thermal behaviour of honeys to detect adulterations effects, that is, sugar syrups and classify honeys (Robinia, Lavender, Chestnut and Fir). Hernandez et al. [180] characterised different types of honey produced in the Canary Islands according to their mineral contents using atomic absorption spectrophotometry. Guo et al. [181] used an open-ended coaxial-line technology and a network analyser at 10–4500 MH to detect sucrose–adulterated honey using a sucrose content sensor where permittivity of different adulterated and pure honeys was measured. Pure honey possessed higher dielectric constant when compared with pure sucrose syrup. Roshan et al. [182] used UV spectroscopy together with chemometric techniques to develop a simple model developed to describe authentication of monofloral Yemeni Sidr honey. Tuberoso et al. [183] assessed colour coordinates of 17 unifloral honeys types from different geographic locations in Europe using spectrophotometric method. Wang et al. [184] determined geographical origin of honey based on fingerprinting and

Techniques	Samples	References
Gravimetric method for ash	22 eucalyptus and citrus honey samples from Brazil	[59]
Differential scanning calorimetry	*Lavandula*, *Robinia* and *Fir* honeys	[179]
Atomic absorption spectrophotometry	116 samples of monofloral and multifloral honeys	[180]
Dielectric properties of honey	Chinese jujube, yellow locust tree and Chinese milk-vetch	[181]
UV spectroscopy	38 honey (13 genuine monofloral Sidr honeys (Ziziphus spinachristi), 14 Sidr honeys, 5 polyfloral and 6 non-Sidr honeys)	[182]
UV–visible spectrophotometer	305 samples from 17 unifloral honey types	[183]
Fingerprinting and barcoding of proteins by MALDI TOF MS	16 honey samples	[184]
Digital image-based flow-batch system	210 honey samples from southwest of the province of Buenos Aires and Argentina	[185]
Multivariate analysis of colour and mineral composition	77 honey samples collected in Spain	[186]
Atomic absorption spectrometry	6 monofloral honeys and 2 multifloral Spanish honeys	[187]
ICP-MS	163 honey samples from 4 types of honey: linden, vitex, rape and acacia	[188]

Table 8.
Other techniques used in honey authentication.

barcoding of proteins in honey by using matrix assisted laser desorption ionisation time-of-flight mass spectrometry (MALDI TOF MS). The protein fingerprints were used to differentiate geographical origins of honey samples produced from different countries and various states of the USA, including Hawaii. Dominguez et al. [185] studied geographic origin classification of honey samples from Argentina by a conventional flow-batch system with a simple webcam to capture digital images. In this technique, analytical information is generated from colour histograms obtained from digital images using different colour models such as red–green–blue (RGB), hue–saturation–brightness (HSB) and Grayscale. Gonzalez-Miret et al. [186] analysed mineral content and colour characteristics of 77 honey samples to classify them following botanical origin. de Alda-Garcilope et al. [187] characterised six monofloral honeys and two multifloral Spanish honeys to their protected designation of origin "Miel de Granada" using their metal content. Chen et al. [188] used inductively coupled plasma mass spectrometry (ICP-MS) and chemometrics data of 12 mineral elements to classify Chinese honeys to their botanical origins. Using simple gravimetric method, Felsner et al. [59] characterised monofloral honeys by its ash content, a parameter that has been associated with floral source of honey samples with the hierarchical design. For analysis of honey classification for authentication purposes, data collected needs strong statistical analysis such as multivariate analysis, regression analysis or chemometrics like principal component analysis (PCA) and linear discriminant analysis (LDA). More advanced techniques include the chemical finger printing technique for indicating a unique pattern.

4. Conclusion

Determination of honey authenticity by its geographical or botanical origins and its purity is the most important criteria to ensure its safety and quality. The older techniques of melissopalinology, for example, may need to be coupled with newer and more advanced techniques to provide higher precision and accuracy of investigation. Strong analytical methods and procedures are necessary for in-depth analysis of data obtained from instrumental measurements for meaningful research in honey authentication. With new knowledge and information of honey origins and authenticity, there rise the need to update current standards of the Codex Alimentarius and the European Union to incorporate newer information and guidelines for standardisation of honey qualities with respect to authenticity. Newer profiles may include components like the aliphatic organic acids, amino acids, volatile components, flavonoids, carbohydrates, phenolic acids and proteins instead of just the sugar profiles. Guidelines towards discrimination botanical and geographical origin may also be implemented. This review provides insights to encourage researchers to further explore novel detection technologies in authentication studies of food materials.

References

[1] Crane E. History of honey. In: Crane E, editor. Honey, A Comprehensive Survey. London: William Heinemann; 1975. pp. 439-488

[2] European Commission. Regulation (EC) No 178/2002 of the European Parliament and of the council of 28 January 2002 laying down the general principles and requirements of food law, establishing the European food safety authority and laying down procedures in matters of food safety. Official Journal of the European Communities. 2002. Vol. OJ L31. pp. 1-24

[3] Zielinski L, Deja S, Jasicka-Misiak I, Kafarski P. Chemometrics as a tool of origin determination of polish monofloral and multifloral honeys. Journal of Agricultural and Food Chemistry. 2014;**62**:2973-2981

[4] Alvarez-Suarez JM, Tulipani S, Diaz D, Estevez Y, Romandini S, Giampieri F, et al. Antioxidant and antimicrobial capacity of several monofloral Cuban honeys and their correlation with color, polyphenol content and other chemical compounds. Food and Chemical Toxicology. 2010; **48**:2490-2499

[5] Aljadi AM, Kamaruddin MY. Evaluation of the phenolic contents and antioxidant capacities of two Malaysian floral honeys. Food Chemistry. 2004;**85**: 513-518

[6] Ferreira ICFR, Aires E, Barreira JCM, Estevinho LM. Antioxidant activity of Portuguese honey samples: Different contributions of the entire honey and phenolic extract. Food Chemistry. 2009; **114**:1438-1443

[7] Ramanauskiene K, Stelmakiene A, Briedis V, Ivanauskas L, Jakstas V. The quantitative analysis of biologically active compounds in Lithuanian honey. Food Chemistry. 2012;**132**:1544-1548

[8] Yao L, Jiang Y, Singanusong R, Datta N, Raymont K. Phenolic acids in Australian *Melaleuca*, *Guioa*, *Lophostemon*, *Banksia* and *Helianthus* honeys and their potential for floral authentication. Food Research International. 2005;**38**:651-658

[9] Belay A, Solomon WK, Bultossa G, Adgaba N, Melaku S. Physicochemical properties of the Harenna forest honey, bale, Ethiopia. Food Chemistry. 2013; **141**:3386-3392

[10] Yap SK, Chin NL, Yusof YA, Chong KY. Quality characteristics of dehydrated raw Kelulut honey. International Journal of Food Properties. 2019;**22**(1):556-571

[11] FAO, Standard for Honey (CODEX STAN 12). *Codex Alimentarius:* Sugars, Cocoa Products and Chocolate and Miscellaneous Products. Rome, Italy: FAO, Vol; 1981 p. 11

[12] Bogdanov S. Nature and origin of the antibacterial substances in honey. LWT- Food Science and Technology. 1997;**30**:748-753

[13] Beretta G, Granata P, Ferrero M, Orioli M, Facino RM. Standardization of antioxidant properties of honey by a combination of spectrophotometric/ fluorimetric assays and chemometrics. Analytica Chimica Acta. 2005;**533**: 185-191

[14] Ouchemoukh S, Louaileche H, Schweitzer P. Physicochemical characteristics and pollen spectrum of some Algerian honeys. Food Control. 2007;**18**:52-58

[15] Van den Berg AJ, Van Den Worm E, Van Ufford HC, Halkes SB, Hoekstra MJ, Beukelman CJ. An *in vitro* examination of the antioxidant and anti-inflammatory properties of buckwheat honey. Journal of Wound Care. 2008; **17**(4):172-174, 176-178

[16] Gomes S, Dias LG, Moreira LL, Rodrigues P, Estevinho L. Physicochemical, microbiological and antimicrobial properties of commercial honeys from Portugal. Food and Chemical Toxicology. 2010;**48**:544-548

[17] Liu JR, Ye YL, Lin TY, Wang YW, Peng CC. Effect of floral sources on the antioxidant, antimicrobial, and anti-inflammatory activities of honeys in Taiwan. Food Chemistry. 2013;**139**: 938-943

[18] Alvarez-Suarez JM, Tulipani S, Romandini S, Bertoli E, Battino M. Contribution of honey in nutrition and human health: A review. Mediterranean Journal of Nutrition and Metabolism. 2010;**3**:15-23

[19] Haron MN, Rahman WFWA, Sulaiman SA, Mohamed M. Tualang honey ameliorates restraint stress-induced impaired pregnancy outcomes in rats. European Journal of Integrative Medicine. 2014;**6**:657-663

[20] Mosavat M, Ooi FK, Mohamed M. Effects of honey supplementation combined with different jumping exercise intensities on bone mass, serum bone metabolism markers and gonadotropins in female rats. BMC Complementary and Alternative Medicine. 2014;**14**:126

[21] Mohamed M, Sulaiman SA, Sirajudeen KNS. Protective effect of honey against cigarette smoke induced-impaired sexual behavior and fertility of male rats. Toxicology and Industrial Health. 2013;**29**:264-271

[22] Tosun M. Detection of adulteration in honey samples added various sugar syrups with $^{13}C/^{12}C$ isotope ratio analysis method. Food Chemistry. 2013; **138**:1629-1632

[23] Bogdanov S, Gallmann P. Authenticity of honey and other bee products state of the art. Animal Production and Dairy Products (ALP) Science. 2008;**520**:1-12

[24] Soares S, Amaral JS, Oliveira MBPP, Mafra I. Improving DNA isolation from honey for the botanical origin identification. Food Control. 2015;**48**: 130-136

[25] Anklam E. A review of the analytical methods to determine the geographical and botanical origin of honey. Food Chemistry. 1998;**63**:549-562

[26] Ferreres F, Giner JM, Tomas-Barberan FA. A comparative study of hesperetin and methyl antranilate as markers of the floral origin of citrus honey. Journal of the Science of Food and Agriculture. 1994;**65**:371-372

[27] Molan PD. In: Ashurst PR, Dennis MJ, editors. Food Authentication. London: Chapman and Hall; 1996

[28] Ouchemoukh S, Schweitzer P, Bey MB, Djoudad-Kadji H, Louaileche H. HPLC sugar profiles of Algerian honeys. Food Chemistry. 2010; **121**:561-568

[29] Chudzinska M, Baralkiewicz D. Estimation of honey authenticity by multielements characteristics using inductively coupled plasma-mass spectrometry (ICP-MS) combined with chemometrics. Food and Chemical Toxicology. 2010;**48**:284-290

[30] Kenjeric D, Mandic ML, Primorac LJ, Bubalo D, Perl A. Flavonoid profile of *Robinia* honeys produced in Croatia. Food Chemistry. 2007;**102**:683-690

[31] Cajka T, Hajslova J, Pudil F, Riddellova K. Traceability of honey origin based on volatiles pattern processing by artificial neural networks. Journal of Chromatography A. 2009; **1216**:1458-1462

[32] Jerkovic I, Marijanovic Z, Kezic J, Gugic M. Headspace, volatile and semi-volatile organic compounds diversity and radical scavenging activity of ultrasonic solvent extracts from *Amorpha fruticosa* honey samples. Molecules. 2009;**14**:2717-2728

[33] Rebane R, Herodes K. Evaluation of the botanical origin of Estonian uni-and polyfloral honeys by amino acid content. Journal of Agricultural and Food Chemistry. 2008;**56**:10716-10720

[34] Stanimirova I, Ustünb B, Cajka T, Riddelova K, Hajslova J, Buydens LMC. Tracing the geographical origin of honeys based on volatile compounds profiles assessment using pattern recognition techniques. Food Chemistry. 2010;**118**:171-176

[35] Verzera A, Tripodi G, Condurso C, Dima G, Marra A. Chiral volatile compounds for the determination of orange honey authenticity. Food Control. 2014;**39**:237-243

[36] Wang J, Kliks MM, Jun S, Jackson M, Li QX. Rapid analysis of glucose, fructose, sucrose, and maltose in honeys from different geographic regions using Fourier transform infrared spectroscopy and multivariate analysis. Journal of Food Science. 2010;**75**: C208-C214

[37] Donarski JA, Jones SA, Charlton AJ. Application of cryoprobe ^{1}H nuclear magnetic resonance spectroscopy and multivariate analysis for the verification of Corsican honey. Journal of Agricultural and Food Chemistry. 2008; **56**:5451-5456

[38] Kropf U, Golob T, Necemer M, Kump P, Korosec M, Bertoncelj J, et al. Carbon and nitrogen natural stable isotopes in Slovene honey: Adulteration and botanical and geographical aspects. Journal of Agricultural and Food Chemistry. 2010;**58**:12794-12803

[39] Schellenberg A, Chmielus S, Schlicht C, Camin F, Perini M, Bontempo L, et al. Multielement stable isotope ratios (H, C, N, S) of honey from different European regions. Food Chemistry. 2010;**121**:770-777

[40] Zakaria A, Md Shakaff AY, Masnan MZ, Ahmad MN, Adom AH, Jaafar MN, et al. A biomimetic sensor for the classification of honeys of different floral origin and the detection of adulteration. Sensors. 2011;**11**(8): 7799-7822

[41] Tahir HE, Xiaobo Z, Xiaowei H, Jiyong S, Mariod AA. Discrimination of honeys using colorimetric sensor arrays, sensory analysis and gas chromatography techniques. Food Chemistry. 2016;**206**:37-43

[42] Bertelli D, Lolli M, Papotti G, Bortolotti L, Serra G, Plessi M. Detection of honey adulteration by sugar syrups using one-dimensional and two-dimensional high-resolution nuclear magnetic resonance. Journal of Agricultural and Food Chemistry. 2010; **58**:8495-8501

[43] Cordella C, Militao JSLT, Clement MC, Drajnudel P, Cabrol-Bass D. Detection and quantification of honey adulteration via direct incorporation of sugar syrups or bee-feeding: Preliminary study using high-performance anion exchange chromatography with pulsed amperometric detection (HPAEC-PAD) and chemometrics. Analytica Chimica Acta. 2005;**531**:239-248

[44] Cotte JF, Casabianca H, Lheritier J, Perrucchietti C, Sanglar C, Waton H, et al. Study and validity of ^{13}C stable carbon isotopic ratio analysis by mass spectrometry and ^{2}H site-specific natural isotopic fractionation by nuclear magnetic resonance isotopic measurements to characterize and control the authenticity of honey.

Analytica Chimica Acta. 2007;**582**: 125-136

[45] de la Fuente E, Sanz ML, Martinez-Castro I, Sanz J, Ruiz-Matute AI. Volatile and carbohydrate composition of rare unifloral honeys from Spain. Food Chemistry. 2007;**105**:84-93

[46] Nozal Nalda MJ, Bernal Yague JL, Diego Calva JC, Martin Gomez MT. Classifying honeys from the Soria Province of Spain via multivariate analysis. Analytical and Bioanalytical Chemistry. 2005;**382**:311-319

[47] de la Fuente E, Ruiz-Matute AI, Valencia-Barrera RM, Sanz J, Martinez Castro I. Carbohydrate composition of Spanish unifloral honeys. Food Chemistry. 2011;**129**:1483-1489

[48] Baroni MV, Nores ML, Diaz MDP, Chiabrando GA, Fassano JP, Costa C, et al. Determination of volatile organic compound patterns characteristic of five unifloral honey by solid-phase microextraction-gas chromatography-mass spectrometry coupled to chemometrics. Journal of Agricultural and Food Chemistry. 2006;**54**:7235-7241

[49] Corbella E, Cozzolino D. Classification of the floral origin of Uruguayan honeys by chemical and physical characteristics combined with chemometrics. LWT - Food Science and Technology. 2006;**39**:534-539

[50] Bertoncelj J, Golob T, Kropf U, Korosec M. Characterisation of Slovenian honeys on the basis of sensory and physicochemical analysis with a chemometric approach. International Journal of Food Science and Technology. 2011;**46**:1661-1671

[51] Bertoncelj J, Polak T, Kropf U, Korosec M, Golob T. LC-DAD-ESI/MS analysis of flavonoids and abscisic acid with chemometric approach for the classification of Slovenian honey. Food Chemistry. 2011;**127**:296-302

[52] Kek SP, Chin NL, Tan SW, Yusof YA, Chua LS. Classification of entomological origin of honey based on its physicochemical and antioxidant properties. International Journal of Food Properties. 2017;**20**:S2723-S2738

[53] Kek SP, Chin NL, Tan SW, Yusof YA, Chua LS. Classification of honey from its bee origin via chemical profiles and mineral content. Food Analytical Methods. 2017;**10**:19-30

[54] Kek SP, Chin NL, Tan SW, Yusof YA, Chua LS. Molecular identification of honey entomological origin based on bee mitochondrial 16S rRNA and COI gene sequences. Food Control. 2017;**78**:150-159

[55] Kek SP, Chin NL, Tan SW, Yusof YA, Chua LS. Comparison of DNA extraction methods for entomological origin identification of honey using simple additive weighting method. International Journal of Food Science and Technology. 2018;**53**: 2490-2499

[56] Lazarevic KB, Andric F, Trifkovic J, Tesic Z, Milojkovic-Opsenica D. Characterisation of Serbian unifloral honeys according to their physicochemical parameters. Food Chemistry. 2012;**132**:2060-2064

[57] Silvano MF, Varela MS, Palacio MA, Ruffinengo S, Yamul DK. Physicochemical parameters and sensory properties of honeys from Buenos Aires region. Food Chemistry. 2014;**152**:500-507

[58] Terrab A, Maria J, Diez MJ, Heredia FJ. Palynological, physico-chemical and colour characterization of Moroccan honeys: III. Other unifloral honey types. International Journal of Food Science and Technology. 2003;**38**: 395-402

[59] Felsner ML, Cano CB, Bruns RE, Watanabe HM, Almeida-Muradian LB,

Matos JR. Characterization of monofloral honeys by ash contents through a hierarchical design. Journal of Food Composition and Analysis. 2004; **17**:737-747

[60] Anupama D, Bhat KK, Sapna VK. Sensory and physico-chemical properties of commercial samples of honey. Food Research International. 2003;**36**:183-191

[61] Kaskoniene V, Venskutonis PR, Ceksteryte V. Carbohydrate composition and electrical conductivity of different origin honeys from Lithuania. LWT - Food Science and Technology. 2010;**43**:801-807

[62] Manzanares AB, Garcia ZH, Galdon BR, Rodriguez ER, Romero CD. Differentiation of blossom and honeydew honeys using multivariate analysis on the physicochemical parameters and sugar composition. Food Chemistry. 2011;**126**:664-672

[63] Kamboj R, Bera MB, Nanda V. Evaluation of physico-chemical properties, trace metal content and antioxidant activity of Indian honeys. International Journal of Food Science and Technology. 2013;**48**:578-587

[64] Stolzenbach S, Byrne DV, Bredie WLP. Sensory local uniqueness of Danish honeys. Food Research International. 2011;**44**:2766-2774

[65] Castro-Vazquez L, Diaz-Maroto MC, Gonzalez-Vinas MA, Perez-Coello MS. Differentiation of monofloral citrus, rosemary, eucalyptus, lavender, thyme and heather honeys based on volatile composition and sensory descriptive analysis. Food Chemistry. 2009;**112**:1022-1030

[66] Castro-Vazquez L, Leon-Ruiz V, Alanon ME, Perez-Coello MS, Gonzalez-Porto AV. Floral origin markers for authenticating Lavandin honey (*Lavandula angustifolia* x *latifolia*). Discrimination from lavender honey (*Lavandula latifolia*). Food Control. 2014;**37**:362-370

[67] Castro-Vazquez L, Diaz-Maroto M, de Torres C, Perez-Coello M. Effect of geographical origin on the chemical and sensory characteristics of chestnut honeys. Food Research International. 2010;**43**:2335-2340

[68] Doner LW, White JW, Phillips JG. Gas-liquid chromatographic test for honey adulteration by high fructose corn syrup. Journal of the Association of Official Analytical Chemists. 1979;**62**: 186-189

[69] Kushnir I. Sensitive thin layer chromatographic detection of high fructose corn sirup and other adulterants in honey. Journal of the Association of Official Analytical Chemists. 1979;**62**:917-920

[70] Lipp J, Ziegler H, Conrady E. Detection of high fructose- and other syrups in honey using high-pressure liquid chromatography. Zeitschrift für Lebensmittel-Untersuchung und -Forschung. 1988;**187**:3342-3338

[71] Prodolliet J, Hischenhuber C. Food authentication by carbohydrate chromatography. Zeitschrift für Lebensmitteluntersuchung und -Forschung A. 1998;**207**:1-12

[72] Morales V, Corzo N, Sanz ML. HPAEC-PAD oligosaccharide analysis to detect adulterations of honey with sugar syrups. Food Chemistry. 2008;**107**: 922-928

[73] Xue X, Wang Q, Li Y, Wu L, Chen L, Zhao J, et al. 2-Acetylfuran-3-glucopyranoside as a novel marker for the detection of honey adulterated with rice syrup. Journal of Agricultural and Food Chemistry. 2013;**61**:7488-7493

[74] Puscas A, Hosu A, Cimpoiu C. Application of a newly developed and

validated high-performance thin-layer chromatographic method to control honey adulteration. Journal of Chromatography A. 2013;**1272**:132-135

[75] Cotte JF, Casabianca H, Chardon S, Lheritier J, Grenier-Loustalot MF. Chromatographic analysis of sugars applied to the characterization of monofloral honey. Analytical and Bioanalytical Chemistry. 2004;**380**: 698-705

[76] Gilbert J, Shephard MJ, Wallwork MA, Harris RG. Determination of the geographical origin of honeys by multivariate analysis of gas chromatographic data on their free amino acid content. Journal of Apicultural Research. 1981; **20**:125-135

[77] Pawlowska M, Armstrong DW. Evaluation of enantiometric purity of selected amino acids in honey. Chirality. 1994;**6**:270-276

[78] Cotte JF, Casabianca H, Giroud B, Albert M, Lheritier J, Grenier-Loustalot MF. Characterization of honey amino acid profiles using high-pressure liquid chromatography to control authenticity. Analytical and Bioanalytical Chemistry. 2004;**378**:1342-1350

[79] Iglesias MT, Martin-Alvarez PJ, Polo MC, de Lorenzo C, Pueyo E. Protein analysis of honeys by fast protein liquid chromatography: Application to differentiate floral and honeydew honeys. Journal of Agricultural and Food Chemistry. 2006; **54**:8322-8327

[80] Keckes J, Trifkovic J, Andric F, Jovetic M, Tesic Z, Milojkovic-Opsenica D. Amino acids profile of Serbian unifloral honeys. Journal of the Science of Food and Agriculture. 2013 **93**: 3368-3376

[81] Wang J, Xue X, Du X, Cheng N, Chen L, Zhao J, et al. Identification of acacia honey adulteration with rape honey using liquid chromatography–electrochemical detection and chemometrics. Food Analytical Methods. 2014;**7**:2003-2012

[82] Yao L, Jiang Y, D'Arcy B, Singanusong R, Datta N, Caffin N, et al. Quantitative high-performance liquid chromatography analyses of flavonoids in Australian *Eucalyptus* honeys. Journal of Agricultural and Food Chemistry. 2004;**52**:210-214

[83] Yao L, Jiang Y, Singanusong R, D'Arcy B, Datta N, Caffin N, et al. Flavonoids in Australian *Melaleuca*, *Guioa*, *Lophostemon*, *Banksia* and *Helianthus* honeys and their potential for floral authentication. Food Research International. 2004;**37**:166-174

[84] Yao L, Jiang Y, Singanusong R, Datta N, Raymont K. Phenolic acids and abscisic acid in Australian *Eucalyptus* honeys and their potential for floral authentication. Food Chemistry. 2004; **86**:169-177

[85] Jasicka-Misiak I, Poliwoda A, Deren M, Kafarski P. Phenolic compounds and abscisic acid as potential markers for the floral origin of two polish unifloral honeys. Food Chemistry. 2012;**131**:1149-1156

[86] Perna A, Intaglietta I, Simonetti A, Gambacorta E. A comparative study on phenolic profile, vitamin C content and antioxidant activity of Italian honeys of different botanical origin. International Journal of Food Science and Technology. 2013;**48**:1899-1908

[87] Karabagias IK, Vavoura MV, Nikolaou C, Badeka A, Kontakos S, Kontominas MG. Floral authentication of Greek unifloral honeys based on the combination of phenolic compounds, physicochemical parameters and chemometrics. Food Research International. 2014;**62**:753-760

[88] Campone L, Piccinelli AL, Pagano I, Carabetta S, Sanzo RD, Russo M, et al. Determination of phenolic compounds in honey using dispersive liquid–liquid microextraction. Journal of Chromatography A. 2014;**1334**:9-15

[89] Kenjeric D, Mandic ML, Primorac L, Cacic F. Flavonoid pattern of sage (*Salvia officinalis* L.) unifloral honey. Food Chemistry. 2008;**110**:187-192

[90] Liang Y, Cao W, Chen W, Xiao X, Zheng J. Simultaneous determination of four phenolic components in citrus honey by high performance liquid chromatography using electrochemical detection. Food Chemistry. 2009;**114**: 1537-1541

[91] Escriche I, Kadar M, Juan-Borras M, Domenech E. Using flavonoids, phenolic compounds and headspace volatile profile for botanical authentication of lemon and orange honeys. Food Research International. 2011;**44**:1504-1513

[92] Escriche I, Kadar M, Juan-Borrás M, Domenech E. Suitability of antioxidant capacity, flavonoids and phenolic acids for floral authentication of honey. Impact of industrial thermal treatment. Food Chemistry. 2014;**142**:135-143

[93] Khalil MI, Alam N, Moniruzzaman M, Sulaiman SA, Gan SH. Phenolic acid composition and antioxidant properties of Malaysian honeys. Journal of Food Science. 2011; **76**:C921-C928

[94] Zhang X-H, Wu H-L, Wang J-Y, Tu D-Z, Kang C, Zhao J, et al. Fast HPLC-DAD quantification of nine polyphenols in honey by using second-order calibration method based on trilinear decomposition algorithm. Food Chemistry. 2013;**138**:62-69

[95] Ceballos L, Pino JA, Quijano-Celis CE, Dago A. Optimization of a HS-SPME/GC-MS method for determination of volatile compounds in some Cuban unifloral honeys. Journal of Food Quality. 2010;**33**:507-528

[96] Plutowska B, Chmiel T, Dymerski T, Wardencki W. A headspace solid-phase microextraction method development and its application in the determination of volatiles in honeys by gas chromatography. Food Chemistry. 2011;**126**:1288-1298

[97] Perez RA, Sanchez-Brunete C, Calvo RM, Tadeo JL. Analysis of volatiles from Spanish honeys by solid-phase microextraction and gas chromatography-mass spectrometry. Journal of Agricultural and Food Chemistry. 2002;**50**:2633-2637

[98] Alissandrakis E, Tarantilis PA, Harizanis PC, Polissiou M. Comparison of the volatile composition in thyme honeys from several origins in Greece. Journal of Agricultural and Food Chemistry. 2007;**55**:8152-8157

[99] Daher S, Gulacar FO. Analysis of phenolic and other aromatic compounds in honeys by solid-phase microextraction followed by gas chromatography-mass spectrometry. Journal of Agricultural and Food Chemistry. 2008;**56**:5775-5780

[100] Senyuva HZ, Gilbert J, Silici S, Charlton A, Dal C, Gurel N, et al. Profiling Turkish honeys to determine authenticity using physical and chemical characteristics. Journal of Agricultural and Food Chemistry. 2009;**57**:3911-3919

[101] Bianchi F, Mangia A, Mattarozzi M, Musci M. Characterization of the volatile profile of thistle honey using headspace solid-phase microextraction and gas chromatography–mass spectrometry. Food Chemistry. 2011;**129**:1030-1036

[102] Soria AC, Gonzalez M, de Lorenzo C, Martinez-Castro I, Sanz J. Characterization of artisanal honeys

from Madrid (Central Spain) on the basis of their melissopalynological, physicochemical and volatile composition data. Food Chemistry. 2004;**85**:121-130

[103] Soria AC, Sanz J, Martinez-Castro I. SPME followed by GC–MS: A powerful technique for qualitative analysis of honey volatiles. European Food Research and Technology. 2009; **228**:579-590

[104] Juan-Borrás M, Domenech E, Hellebrandova M, Escriche I. Effect of country origin on physicochemical, sugar and volatile composition of acacia, sunflower and tilia honeys. Food Research International. 2014;**60**:86-94

[105] Aliferis KA, Tarantilis PA, Harizanis PC, Alissandrakis E. Botanical discrimination and classification of honey samples applying gas chromatography/mass spectrometry fingerprinting of headspace volatile compounds. Food Chemistry. 2010;**121**: 856-862

[106] Spanik I, Pazitna A, Siska P, Szolcsanyi P. The determination of botanical origin of honeys based on enantiomer distribution of chiral volatile organic compounds. Food Chemistry. 2014;**158**:497-503

[107] Ruiz-Matute AI, Soria AC, Martinez-Castro I, Sanz ML. A new methodology based on GC-MS to detect honey adulteration with commercial syrups. Journal of Agricultural and Food Chemistry. 2007;**55**:7264-7269

[108] Ruiz-Matute AI, Rodriguez-Sanchez S, Sanz ML, Martinez-Castro I. Detection of adulterations of honey with high fructose syrups from inulin by GC analysis. Journal of Food Composition and Analysis. 2010;**23**:273-276

[109] Langford V, Gray J, Foulkes B, Bray P, McEwan MJ. Application of selected ion flow tube-mass spectrometry to the characterization of monofloral New Zealand honeys. Journal of Agricultural and Food Chemistry. 2012;**60**:6806-6815

[110] Agila A, Barringer S. Application of selected ion flow tube mass spectrometry coupled with chemometrics to study the effect of location and botanical origin on volatile profile of unifloral American honeys. Journal of Food Science. 2012;**77**: C1103-C1108

[111] Tuberoso CIG, Bifulco E, Caboni P, Cottiglia F, Cabras P, Floris I. Floral markers of strawberry tree (*Arbutus unedo* L.) honey. Journal of Agricultural and Food Chemistry. 2010;**58**:384-389

[112] Zhou J, Yao L, Li Y, Chen L, Wu L, Zhao J. Floral classification of honey using liquid chromatography–diode array detection–tandem mass spectrometry and chemometric analysis. Food Chemistry. 2014;**145**:941-949

[113] Oelschlaegel S, Gruner M, Wang P-N, Boettcher A, Koelling-Speer I, Speer K. Classification and characterization of manuka honeys based on phenolic compounds and methylglyoxal. Journal of Agricultural and Food Chemistry. 2012;**60**:7229-7237

[114] Trautvetter S, Koelling-Speer I, Speer K. Confirmation of phenolic acids and flavonoids in honeys by UPLC-MS. Apidologie. 2009;**40**:140-150

[115] Martin IG, Macias EM, Sanchez JS, Rivera BG. Detection of honey adulteration with beet sugar using stable isotope methodology. Food Chemistry. 1998;**61**:281-286

[116] Padovan GJ, De Jong D, Rodrigues LP, Marchini JS. Detection of adulteration of commercial honey samples by the $^{13}C/^{12}C$ isotopic ratio. Food Chemistry. 2003;**82**:633-636

[117] Marini F, Magri AL, Balestrieri F, Fabretti F, Marini D. Supervised pattern

recognition applied to the discrimination of the floral origin of six types of Italian honey samples. Analytica Chimica Acta. 2004;**515**:117-125

[118] Cinar SB, Eksi A, Coskun I. Carbon isotope ratio ($^{13}C/^{12}C$) of pine honey and detection of HFCS adulteration. Food Chemistry. 2014;**157**:10-13

[119] Cabanero AI, Recio JL, Ruperez M. Liquid chromatography coupled to isotope ratio mass spectrometry: A new perspective on honey adulteration detection. Journal of Agricultural and Food Chemistry. 2006;**54**:9719-9727

[120] Cengiz MF, Durak MZ, Ozturk M. In-house validation for the determination of honey adulteration with plant sugars (C4) by isotope ratio mass spectrometry (IR-MS). LWT - Food Science and Technology. 2014;**57**:9-15

[121] Daniele G, Maitre D, Casabianca H. Identification, quantification and carbon stable isotopes determinations of organic acids in monofloral honeys. A powerful tool for botanical and authenticity control. Rapid Communications in Mass Spectrometry. 2012;**26**:1993-1998

[122] Elflein L, Raezke K-P. Improved detection of honey adulteration by measuring differences between $^{13}C/^{12}C$ stable carbon isotope ratios of protein and sugar compounds with a combination of elemental analyzer - isotope ratio mass spectrometry and liquid chromatography - isotope ratio mass spectrometry ($\delta^{13}C$-EA/LC-IRMS). Apidologie. 2008;**39**:574-587

[123] Simsek A, Bilsel M, Goren AC. $^{13}C/^{12}C$ pattern of honey from Turkey and determination of adulteration in commercially available honey samples using EA-IRMS. Food Chemistry. 2012; **130**:1115-1121

[124] Chen H, Fan C, Wang Z, Chang Q, Wang W, Li X, et al. Evaluation of measurement uncertainty in EA–IRMS: For determination of $\delta^{13}C$ value and C-4 plant sugar content in adulterated honey. Accreditation and Quality Assurance. 2013;**18**:351-358

[125] Frew R, McComb K, Croudis L, Clark D, Van Hale R. Modified sugar adulteration test applied to New Zealand honey. Food Chemistry. 2013; **141**:4127-4131

[126] Ruoff K, Luginbuhl W, Bogdanov S, Bosset JO, Estermann B, Ziolko T, et al. Authentication of the botanical origin of honey by near-infrared spectroscopy. Journal of Agricultural and Food Chemistry. 2006; **54**:6867-6872

[127] Ruoff K, Luginbuhl W, Kunzli R, Bogdanov S, Bosset JO, von der Ohe K, et al. Authentication of the botanical and geographical origin of honey by front-face fluorescence spectroscopy. Journal of Agricultural and Food Chemistry. 2006;**54**:6858-6866

[128] Ruoff K, Werner Luginbuhl W, Bogdanov S, Bosset JO, Estermann B, Ziolko T, et al. Quantitative determination of physical and chemical measurands in honey by near-infrared spectrometry. European Food Research and Technology. 2007;**225**:415-423

[129] Ruoff K, Iglesias MT, Luginbuhl W, Bosset J-O, Bogdanov S, Amado R. Quantitative analysis of physical and chemical measurands in honey by mid-infrared spectrometry. European Food Research and Technology. 2006;**223**:22-29

[130] Dvash L, Afik O, Shafir S, Schaffer A, Yeselson Y, Dag A, et al. Determination by near-infrared spectroscopy of perseitol used as a marker for the botanical origin of avocado (*Persea americana* mill.) honey. Journal of Agricultural and Food Chemistry. 2002;**50**:5283-5287

[131] Qiu PY, Ding HB, Tang YK, Xu RJ. Determination of chemical composition of commercial honey by near-infrared spectroscopy. Journal of Agricultural and Food Chemistry. 1999;**47**:2760-2765

[132] Woodcock T, Downey G, O'Donnell CP. Near infrared spectral fingerprinting for confirmation of claimed PDO provenance of honey. Food Chemistry. 2009;**114**:742-746

[133] Woodcock T, Downey G, Kelly JD, O'Donnell C. Geographical classification of honey samples by near-infrared spectroscopy: A feasibility study. Journal of Agricultural and Food Chemistry. 2007;**55**:9128-9134

[134] Liang XY, Li XY, Wu WJ. Classification of floral origins of honey by NIR and chemometrics. Advanced Materials Research. 2013;**605–607**: 905-909

[135] Latorre CH, Pena RM, Garcia-Martin S, Garcia B. A fast chemometric procedure based on NIR data for authentication of honey with protected geographical indication. Food Chemistry. 2013;**141**:3559-3565

[136] Etzold E, Lichtenberg-Kraag B. Determination of the botanical origin of honey by Fourier-transformed infrared spectroscopy: An approach for routine analysis. European Food Research and Technology A. 2008;**227**:579-586

[137] Sivakesava S, Irudayaraj J. Classification of simple and complex sugar adulterants in honey by mid-infrared spectroscopy. International Journal of Food Science and Technology. 2002;**37**:351-360

[138] Downey G, Fouratier V, Kelly JD. Detection of honey adulteration by addition of fructose and glucose using near infrared transflectance spectroscopy. Journal of Near Infrared Spectroscopy. 2003;**11**:447-456

[139] Zhu X, Li S, Shan Y, Zhang Z, Li G, Su D. Detection of adulterants such as sweeteners materials in honey using near-infrared spectroscopy and chemometrics. Journal of Food Engineering. 2010;**101**:92-97

[140] Chen L, Xue X, Ye Z, Zhou J, Chen F, Zhao J. Determination of Chinese honey adulterated with high fructose corn syrup by near infrared spectroscopy. Food Chemistry. 2011; **128**:1110-1114

[141] Li S, Shan Y, Zhu X, Zhang X, Ling G. Detection of honey adulteration by high fructose corn syrup and maltose syrup using Raman spectroscopy. Journal of Food Composition and Analysis. 2012;**28**:69-74

[142] Paradkar MM, Irudayaraj J. Discrimination and classification of beet and cane inverts in honey by FT-Raman spectroscopy. Food Chemistry. 2001;**76**: 231-239

[143] Pierna JA, Abbas O, Dardenne P, Baeten V. Discrimination of Corsican honey by FT-Raman spectroscopy and chemometrics. Biotechnology, Agronomy, Society and Environment. 2011;**15**:75-84

[144] Cubero-Leon E, Penalver R, Maquet A. Review on metabolomics for food authentication. Food Research International. 2014;**60**:95-107

[145] de la Mata P, Dominguez-Vidal A, Bosque-Sendra JM, Ruiz-Medina A, Cuadros-Rodríguez L, Ayora-Cañada MJ. Olive oil assessment in edible oil blends by means of ATR-FTIR and chemometrics. Food Control. 2012; **23**:449-455

[146] Kelly JFD, Downey G, Fouratier V. Initial study of honey adulteration by sugar solutions using mid-infrared (MIR) spectroscopy and chemometrics. Journal of Agricultural and Food Chemistry. 2004;**52**:33-39

[147] Kelly JD, Petisco C, Downey G. Application of Fourier transform mid-infrared spectroscopy to the discrimination between Irish artisanal honey and such honey adulterated with various sugar syrups. Journal of Agricultural and Food Chemistry. 2006; **54**:6166-6171

[148] Hennessy S, Downey G, O'Donnell C. Multivariate analysis of attenuated total reflection Fourier transform infrared spectroscopic data to confirm the origin of honeys. Applied Spectroscopy. 2008;**62**:1115-1123

[149] Gallardo-Velázquez T, Osorio-Revilla G, Loa MZ, Rivera-Espinoza Y. Application of FTIR-HATR spectroscopy and multivariate analysis to the quantification of adulterants in Mexican honeys. Food Research International. 2009;**42**:313-318

[150] Fotakis C, Kokkotou K, Zoumpoulakis P, Zervou M. NMR metabolite fingerprinting in grape derived products: An overview. Food Research International. 2013;**54**(1): 1184-1194

[151] Marcone MF, Wang S, Albabish W, Nie S, Somnarain D, Hill A. Diverse food-based applications of nuclear magnetic resonance (NMR) technology. Food Research International. 2013;**51**: 729-747

[152] Consonni R, Cagliani LR. Geographical characterization of polyfloral and acacia honeys by nuclear magnetic resonance and chemometrics. Journal of Agricultural and Food Chemistry. 2008;**56**:6873-6880

[153] Consonni R, Cagliani LR, Cogliati C. Geographical discrimination of honeys by saccharides analysis. Food Control. 2013;**32**:543-548

[154] Boffo EF, Tavares LA, Tobias ACT, Ferreira MMC, Ferreira AG. Identification of components of Brazilian honey by ^{1}H NMR and classification of its botanical origin by chemometric methods. LWT - Food Science and Technology. 2012;**49**:55-63

[155] Donarski JA, Jones SA, Harrison M, Driffield M, Charlton AJ. Identification of botanical biomarkers found in Corsican honey. Food Chemistry. 2010; **118**:987-994

[156] Simova S, Atanassov A, Shishiniova M, Bankova V. A rapid differentiation between oak honeydew honey and nectar and other honeydew honeys by NMR spectroscopy. Food Chemistry. 2012;**134**:1706-1710

[157] Beretta G, Caneva E, Regazzoni L, Bakhtyari NG, Maffei Facino R. A solid-phase extraction procedure coupled to ^{1}H NMR, with chemometric analysis, to seek reliable markers of the botanical origin of honey. Analytica Chimica Acta. 2008;**620**:176-182

[158] Lolli M, Bertelli D, Plessi M, Sabatini AG, Restani C. Classification of Italian honeys by 2D HR-NMR. Journal of Agricultural and Food Chemistry. 2008;**56**(4):1298-1304

[159] Schievano E, Peggion E, Mammi S. ^{1}H nuclear magnetic resonance spectra of chloroform extracts of honey for chemometric determination of its botanical origin. Journal of Agricultural and Food Chemistry. 2010;**58**:57-65

[160] Schievano E, Stocchero M, Morelato E, Facchin C, Mammi S. An NMR-based metabolomic approach to identify the botanical origin of honey. Metabolomics. 2012;**8**:679-690

[161] Ribeiro ROR, Marsico ET, Carneiro CS, Monteiro MLG, Junior CAC, Mano S, et al. Classification of Brazilian honeys by physical and chemical analytical methods and low field nuclear magnetic resonance (LF ^{1}H NMR). LWT - Food Science and Technology. 2014;**55**:90-95

[162] Ribeiro ROR, Marsico ET, Carneiro CS, Monteiro MLG, Junior CC, de Jesus EFO. Detection of honey adulteration of high fructose corn syrup by low field nuclear magnetic resonance (LF ^{1}H NMR). Journal of Food Engineering. 2014;**135**:39-43

[163] White JW. The protein content of honey. Journal of Apicultural Research. 1978;**17**:234-238

[164] Lee CY, Smith NL, Kime RW, Morse RA. Source of the honey protein responsible for apple juice clarification. Journal of Apicultural Research. 1985; **24**:190-194

[165] Stadelmeier M, Bergner KG. The proteins of honey. 7. Behaviour and origin of honey amylase. Zeitschrift für Lebensmittel-Untersuchung und Forschung. 1986;**182**:196-199

[166] Marshall T, Williams KM. Electrophoresis of honey: Characterization of trace proteins from complex biological matrix by silver staining. Analytical Biochemistry. 1987; **167**:301-303

[167] Lee DC, Lee SY, Cha SH, Choi YS, Rhee HI. Discrimination of native bee-honey and foreign bee-honey by SDS-PAGE. Korean Journal of Food Science and Technology. 1998;**30**:1-5

[168] Baroni MV, Chiabrando GA, Costa C, Wunderlin DA. Assessment of the floral origin of honey by SDS-PAGE immunoblot techniques. Journal of Agricultural and Food Chemistry. 2002; **50**:1362-1367

[169] Won S-R, Lee C-Y, Kim J-W, Rhee H-I. Immunological characterization of honey major protein and its application. Food Chemistry. 2009;**113**:1334-1338

[170] Simuth J, Bilikova K, Kovacova E, Kuzmova Z, Schroder W. Immunochemical approach to detection of adulteration in honey: Physiologically active royal jelly protein stimulating TNF-α release is a regular component of honey. Journal of Agricultural and Food Chemistry. 2004;**52**:2154-2158

[171] Bilikova K, Simuth J. New criterion for evaluation of honey: Quantification of Royal Jelly Protein Apalbumin 1 in honey by ELISA. Journal of Agricultural and Food Chemistry. 2010;**58**:8776-8781

[172] Valentini A, Miquel C, Taberlet P. DNA barcoding for honey biodiversity. Diversity. 2010;**2**:610-617

[173] Laube I, Hird H, Brodmann P, Ullmann S, Schone-Michling M, Chisholm J, et al. Development of primer and probe sets for the detection of plant species in honey. Food Chemistry. 2012;**118**:979-986

[174] Olivieri C, Marota I, Rollo F, Luciani S. Tracking plant, fungal, and bacterial DNA in honey specimens. Journal of Forensic Sciences. 2012;**57**: 222-227

[175] Guertler P, Eicheldinger A, Muschler P, Goerlich O, Busch U. Automated DNA extraction from pollen in honey. Food Chemistry. 2014;**149**: 302-306

[176] Jain SA, de Jesus FT, Marchioro GM, de Araujo ED. Extraction of DNA from honey and its amplification by PCR for botanical identification. Food Science and Technology. 2013;**33**:753-756

[177] Waiblinger H-U, Ohmenhaeuser M, Meissner S, Schillinger M, Pietsch K, Goerlich O, et al. In-house and interlaboratory validation of a method for the extraction of DNA from pollen in honey. Journal für Verbraucherschutz und Lebensmittelsicherheit. 2012;**7**:243-254

[178] Cordella C, Antinelli J-F, Aurieres C, Faucon J-P, Cabrol-Bass D,

Sbirrazzuoli N. Use of differential scanning calorimetry (DSC) as a new technique for detection of adulteration in honeys. 1. Study of adulteration effect on honey thermal behavior. Journal of Agricultural and Food Chemistry. 2002; **50**:203-208

[179] Cordella C, Faucon J-P, Cabrol-Bass D, Sbirrazzuoli N. Application of DSC as a tool for honey floral species characterization and adulteration detection. Journal of Thermal Analysis and Calorimetry. 2003;**71**:279-290

[180] Hernandez OM, Fraga JMG, Jimenez AI, Jimenez F, Arias JJ. Characterization of honey from the Canary Islands: Determination of the mineral content by atomic absorption spectrophotometry. Food Chemistry. 2005;**93**:449-458

[181] Guo W, Liu Y, Zhu X, Wanga S. Dielectric properties of honey adulterated with sucrose syrup. Journal of Food Engineering. 2011;**107**:1-7

[182] Roshan A-RA, Gad HA, El-Ahmady SH, Khanbash MS, Abou-Shoer MI, Al-Azizi MM. Authentication of monofloral Yemeni Sidr honey using ultraviolet spectroscopy and chemometric analysis. Journal of Agricultural and Food Chemistry. 2013; **61**:7722-7729

[183] Tuberoso CIG, Jerkovic I, Sarais G, Congiu F, Marijanovic Z, Kus PM. Color evaluation of seventeen European unifloral honey types by means of spectrophotometrically determined CIE $L^*C^*_{ab}h^\circ_{ab}$ chromaticity coordinates. Food Chemistry. 2014;**145**: 284-291

[184] Wang J, Kliks MM, Qu W, Jun S, Shi G, Li QX. Rapid determination of the geographical origin of honey based on protein fingerprinting and barcoding using MALDI TOF MS. Journal of Agricultural and Food Chemistry. 2009; **57**:10081-10088

[185] Dominguez MA, Diniz PHGD, Nezio MSD, de Araujo MCU, Centurion ME. Geographical origin classification of Argentinean honeys using a digital image-based flow-batch system. Microchemical Journal. 2014; **112**:104-108

[186] Gonzalez-Miret ML, Terrab A, Hernanz D, Fernandez-Recamales MA, Heredia FJ. Multivariate correlation between color and mineral composition of honeys and by their botanical origin. Journal of Agricultural and Food Chemistry. 2005;**53**:2574-2580

[187] de Alda-Garcilope C, Gallego-Pico A, Bravo-Yague JC, Garcinuño-Martínez RM, Fernández-Hernando P. Characterization of Spanish honeys with protected designation of origin “Miel de Granada” according to their mineral content. Food Chemistry. 2012;**135**: 1785-1788

[188] Chen H, Fan C, Chang Q, Pang G, Hu X, Lu M, et al. Chemometric determination of the botanical origin for Chinese honeys on the basis of mineral elements determined by ICP-MS. Journal of Agricultural and Food Chemistry. 2014;**62**:2443-2448

Index

A

Agricultural crops 1
Alloxan 25
Aloe vera 26
Alto Balsas Michoacano 5
Amino acids 1,2,37,44,53,71
Antiapoptotic proteins 26
Antibacterial 2
Anticancer 25
Anticancer properties 2
Antidiabetic agent 24
Antifungal 2
Anti-inflammatory action 21
Anti-inflammatory agent 23
Anti-inflammatory2,17,20
Antioxidant action 21
Antioxidant activity 21
Antioxidant effect 21
Antioxidant2,17,21,28,31
Antiviral 17
Apalbumin-1 26
Apigenin 21
Apis mellifera 1,2,49,53,68
Atopic dermatitis (AD) 23

B

B60 lymphocytes 22
Beekeeping 1
Bees' honey 6
Bioactive compounds 17
Botanical 2
Botanical origin 2,7
Botanical origin of honey 5

C

Cancer cell lines 26
Cancer cells 25
Carbohydrates 1,2,33,54,61,71
Cell cultures 20
Cell cycle 26
Cellular proliferation 25
Clinical conditions 22
CodexAlimentarius 2
Colon cancer cell lines 25
Conventional medications 17
C-reactive protein 25
Crucial effects 19
Cultivated plants 2
Cyclooxygenase-1 20

D

Dextrose 25
Diabetes mellitus 24
Diabetic diet 24
Diarrhea 17,22
Differential pulse polarography(DPP) 7
Duodenal ulcers 23

E

Ecological issues 6
Economic and social burdens 24
Egyptian papyri 18
Enterobacter aerogenes 19
Enterococcus 20
Enteropathogenic 22
Entomological source 17
Epithelial cells 20
Etiology 24

Exotoxins 23
Extra floral honey 1

F

Fatty acids 1
Flavonoids 2
Floral honey 1
Food frying 25
Food product1,2,53,66
Frieseomelitta nigra 7
Fungistatic 17

G

Gastric ulcer 17
Geographical origin 2
Geotrigona acapulconis 6,7
Glioma 26
Global concern 2
Glucose-oxidase enzyme 19
Glutathione 21

H

Helicobacterpylori 23
Hellwegeri 7
Hesperetin 21
Homocysteine 25
Honey daily 22
Honey enzymes 2
Honey flora 2
Honey market 2
Honey properties 1
Honey quality 5
Hydrogenperoxide (H2O2) 19
Hypertriglyceridemia 25
Hypoglycemic effect 25

I

Immune booster 17
Immunomodulation 20
Induciblenitric oxide synthase (iNOS) 20
Infectious diseases 17
International level 3

K

Keratinocytes 23

L

Lactobacillus 19
Lavandula 2
Leptospermum polygalifolium 19
Leptospermum scoparium 19
Low environmental impact 1

M

M. fasciata 7
Manuka honey 22
Melipona fasciata 7
Meliponicultural 1
Meliponiculture 5
Meliponine bees 1
Meliponini 5
Melissopalynological 7
Melissopalynological analysis 5
Methicillin-resistant S. aureus (MRSA) 19
Minimum inhibitory concentration (MIC) 20
Mitochondrial membrane 25
MM6 cells 22
Monosaccharaides 24
Multifloral honey 2
Mutagenicity 26

N

N. perilampoides 10
Nannotrigona per lampiodes 7
Natural raw honey 24
Neutrophils 22
Non-peroxide honeys 19
Nutrients 1

O

Oral rehydration solution (ORS) 22
Ornithine decarboxylase 20
Osmolarity 19,28
Oxidative reactions 21
Oxygen radical absorbancecapacity (ORAC) 21

P

P. fulvopilosa 7
Pathogenic microbes 19
Perilampoides 10
Phenolic acids 2,19,71,77,78
Phenomena 6,35
Physicochemical 6
Physiological impacts 21
Pine-oakforests 7
Pinocembrin 21
Plebeia fulvopilosa 6,7
Pollination 1
Pollination process 1
Polliniferous tree 10
Pot-honey characteristics 7
Principal component analysis (PCA) 7
Proapoptotic proteins 26
Programmed cell 25
Prostaglandin E2 20
Pseudomonas aeruginosa 20
Psoriasis 24

Q

Quercetin 21

R

Reactive oxygen species (ROS) 21
Roasting processes 25

S

S. hellwegeri 10
Salmonella 22
Scaptotrigona hellwegeri 6,7,8,9
Short-chain fatty acid (SCFA) 20
Skin fungal infections 24
Skin infection 17
Skin staphylococci 24
Staphylococcus aureus 24
Stingless bees 1,5,6,11,31
Storage condition 17
Streptozotocin 25
Sunburn healing 17

T

T Lymphocytes 22
Tepalcatepec 6
The honeys' antioxidant 21
Therapeutic 2
Therapeutic agent 17
ThromboxaneB2 20
Tissue cultures 25
Tocopherols 21
Tropical dry forest 7
Tumor cell 26
Tyrosine kinase 20

U

Unifloral 2
Unifloral honey 2

V

Vancomycin-resistant enterococci (VRE) 19
Volatile compounds 2
Volatile oil of propolis (VOP) 21

W

Wound healing 17,26,30,48